NEWTON'S APPLE

AND OTHER MYTHS ABOUT SCIENCE

NEWTON'S APPLE

AND OTHER MYTHS ABOUT SCIENCE

Edited by Ronald L. Numbers
and Kostas Kampourakis

HARVARD UNIVERSITY PRESS

Cambridge, Massachusetts & London, England

2015

Library of Congress Cataloging-in-Publication Data

Newton's apple and other myths about science / edited by Ronald L. Numbers and
Kostas Kampourakis.
    pages cm
  Includes bibliographical references and index.
  ISBN 978-0-674-96798-4 (cloth)
    1. Errors, Scientific—Popular works.   2. Errors, Scientific—History—Popular
works.   3. Science—History—Popular works.   I. Numbers, Ronald L., editor.
II. Kampourakis, Kostas, editor.
  Q172.5.E77N49 2015
  001.96—dc23   2015014096

To

Nicolaas Rupke and his colleagues at Washington
and Lee University for hosting a wonderful
conference leading to this book in May 2014

CONTENTS

II. NINETEENTH CENTURY

IV. GENERALIZATIONS

ACKNOWLEDGMENTS

This book was conceived in autumn 2009, when Kostas Kampourakis read and became inspired by Ronald Numbers's *Galileo Goes to Jail and Other Myths about Science and Religion*, and immediately imagined a similar book devoted primarily to myths abounding in science education. The two met for the first time at the Darwin Now Conference in Alexandria, Egypt, in November of that year, and Kostas asked Ron the secret to the quality of *Galileo Goes to Jail*. Ron gladly shared the secret: "I asked experts to write on each topic." Kostas kept that secret in mind and in July 2012 paid a visit to Madison, Wisconsin, to propose to Ron to work together on a sequel to *Galileo Goes to Jail*, focusing on historical myths about science. Thus, a deal was sealed.

Crucial to the success of this project has been the collaboration of over two dozen colleagues, but none more so than Nicolaas Rupke, a longtime friend of Numbers, who invited all of us to a working conference at Washington and Lee University, May 9–10, 2014, supported financially by the Johnson Lecture Series, the Dean's Office, and the Center for International Learning. Rupke and his colleagues—President Kenneth Ruscio, Provost Daniel Wubah, Dean Suzanne Keen, Mark Rush, Gregory Cooper, and Laurent Boetsch—treated us royally, as did Carolyn Wingrove-Thomas and Alicia Shires. Besides the contributors to this book,

Richard Burian (Virginia Polytechnic Institute and State University) and Gregory Macklem (University of Notre Dame) provided insightful commentary and criticism.

Joining us at the conference was Michael Fisher, executive editor for science and medicine at Harvard University Press, who gave us advice and encouragement. We are grateful to him, as well as to Andrew Kinney and Lauren Esdaile from Harvard University Press, and Deborah Grahame-Smith and Jamie Thaman from Westchester Publishing Services, for their work during the production of the book. This book was right from the start intended to be a sequel to *Galileo Goes to Jail*, and it is therefore a pleasure to have it published by Harvard University Press.

Kostas Kampourakis would like to thank Ron Numbers for accepting to work with him on a book Ron could have easily edited on his own. The present book would never have existed without Ron's competence, experience, open-mindedness, and sense of humor, which make him the best coeditor one could wish for. Kostas would also like to thank his family for their love and support. Ron thanks Kostas for his vision, importunity, and dedication; and Margie Wilsman, his favorite science educator, for her inspiration and affection. Both Kostas and Ron thank the contributors to this volume for their cooperation, promptness, and high-quality essays, which made possible the completion of the book manuscript within a year of sending out the invitations to contribute.

NEWTON'S APPLE

AND OTHER MYTHS ABOUT SCIENCE

INTRODUCTION

Ronald L. Numbers and Kostas Kampourakis

Great is the power of steady misrepresentation; but the history of science shows that fortunately this power does not long endure.

—Charles Darwin, *On the Origin of Species by Means of Natural Selection* (1872)

"Who cares?" a critical reader of this book might ask. Who cares about Newton's apple or Mendel's peas? Why should anyone care to learn more about the historical episodes and ideas discussed in this book? Perhaps a biologist should know more about Darwin or Mendel, a physicist about Newton and Einstein, a chemist about Wöhler and Pauling, and so on. But maybe not? Perhaps even science students and scientists should not worry too much about learning the details of the life and work of the giants of their discipline. In any case, these giants are long dead, and their theories have changed or disappeared. Contemporary science is very different from what "men of science" used to do in the past. In fact, about half of the historical figures in this book were involved in natural history or natural philosophy, rather than in what we now call science. Therefore, why bother to know the details of what seem to be stories that are esoteric to specific disciplines?

The answer to the reasonable question Who cares? is simple and clear but not always explicit and straightforward: one should care because historical myths about science hinder science literacy and advance a distorted portrayal of how science has been—and is—done. Contrary to what Charles Darwin wrote in the opening quotation to this introduction, the history of science unfortunately shows that the power of misrepresentation endures, and as a consequence, myths remain widespread. We should perhaps note at this point that as in *Galileo Goes to Jail and Other Myths about Science and Religion,* the model for this book, we do not employ the term "myth" in any sophisticated academic sense but rather in the way it is used in everyday conversation—to designate a claim that is false.[1]

The public learns about science in formal (e.g., schools), nonformal (e.g., museums), and informal (e.g., mass media) ways. In all cases, alongside the content knowledge they acquire about a specific discipline (such as Newtonian mechanics in school, evolution in a natural-history museum, or the genetic basis of a disease in the news), people also get an implicit message about how science was done in each case. This message is often transmitted through a narrative about how a scientist "discovered" what students now learn as a "fact." For instance, it is customary to read in a newspaper about some scientist in a university or research center who made a groundbreaking discovery, which has uncovered or is expected to uncover the secrets of a particular natural phenomenon. The implicit emphasis in such accounts is often on how bright that person was, how many years he or she devoted to the respective research, and how important the achievement is.

No one questions the need to be bright and hardworking in order to achieve something significant in science, but that is not the whole story. Traditional narratives often mask other important components of these achievements, such as the contributions of associates and assistants or the possibility that luck may have played a role. Stories that focus on one component of a scientific achievement may ignore some other equally important ones. This can lead to stereotypes about science—some of which are exposed in the last chapters of this book, which focus on how science is practiced and what kind of knowledge it produces. The first chapters, in contrast, explore some clichés about early science and misconceptions about the methods and accomplishments of some well-known scientists.

Students, educators, and general readers need not only to acquire a knowledge of scientific content but also to understand what is called "the nature of science": how science is done, what kinds of questions scientists ask, and what kinds of knowledge they produce. Such scientifically literate citizens will possess a more authentic view of science and will be better able to understand the strengths and limitations of science—and thus make informed decisions about important issues, such as climate change, genetic testing, and biological evolution. Overall, the chapters in this book debunk three kinds of myths: those about the precursors of modern science, those about how science is done, and those about scientists themselves.

Other scholars have single-handedly attempted to correct many of the most egregious myths about science, with greater or lesser success.[2] Too timid to take on this task by ourselves, we solicited the assistance of some twenty-six other experts in the history of science and science education. Together, we episodically or thematically cover the past two thousand years of history. Many of our contributors rank among the leading scholars in the world in their fields of expertise; all of them are experts in their assigned topics. Although some mistakes may have escaped our collective scrutiny, we hope that we have kept them to a minimum.

I

MEDIEVAL AND EARLY MODERN SCIENCE

THAT THERE WAS NO SCIENTIFIC ACTIVITY
BETWEEN GREEK ANTIQUITY AND THE
SCIENTIFIC REVOLUTION

Michael H. Shank

Had Christianity not interrupted the intellectual advance
of mankind and put the progress of science on hold for a
thousand years, the Scientific Revolution might have occurred
a thousand years ago, and our science and technology today
would be a thousand years more advanced.

    —Richard Carrier, "Christianity Was Not Responsible for
Modern Science" (2009)

The widespread myth that there was no scientific activity between
Greek antiquity and the Scientific Revolution is becoming increasingly graphic. There even exists a chart that shows an imagined
hole in the exponential advancement of science left by the Dark
Ages. It assumes that once started, science grows on its own (exponentially here) unless impeded by malevolent forces. Deviations
from expectations thus trigger a search for culprits. Jim Walker,
who constructed the chart, disarmingly writes: "Unfortunately I
do not have the complete database of historical scientific advances
but historians could certainly compile the known scientific advances and even come up with estimated numbers and plug them
into a graph. I suspect the scientific Dark Ages will become even
more apparent and dark."[1] A chart more permanent and damning
than this one recently appeared under the Springer imprimatur,

accompanied by Carl Sagan's assessment of medieval science as "the millennium gap in the middle of the diagram [from Thales (ca. 624–ca. 546 BCE) to 1980, with nothing between Hypatia (ca. 350–415) and Leonardo da Vinci (1452–1519), which] represents a poignant lost opportunity for the human species."[2] When Sagan published his *Cosmos* in 1980, his quip was more than two generations out of date; the 2012 republication now makes it three.

The perpetuation of the myth typically trusts an "authority" in one area (here, Sagan in astronomy) to speak authoritatively about another area (for example, the history of science), about which he repeats out-of-date popular prejudices about one thousand years of medieval stagnation.[3] Obviously no scholar should waste time researching and writing about a period in which nothing happened. Without looking, we therefore know that nothing new can possibly have been discovered or written on the subject. This conclusion is fully consistent with Sagan's statements, thus confirming the latter's "reliable" authority. Ironically, this is just the sort of behavior that is imputed to those stupid medieval folks.

For historians of medieval science, breaking into such vicious circles is a perverse variant on the frustrating game of Whac-A-Mole, in which a player equipped with a mallet scores by hitting on the head moles that randomly pop out of their holes. No sooner has one whacked the dead horse of medieval scientific nothingness back into its grave than it pops out of the ground in new trappings for another wild gallop through popular culture. Few notice the stench.

In the last several years, the myth has enjoyed some glamorous and high-profile vehicles. In 2009, Alejandro Amenábar's beautiful movie *Agora* used Sagan's stunningly anachronistic history of science to make the fifth-century murder of Hypatia in Alexandria the beginning of the Dark Ages. Stephen Greenblatt's *The Swerve* (2011) won the Pulitzer and other prestigious prizes for its fanciful notion that atom-bashing Christianity suppressed Lucretius's *De Rerum Natura* until its recovery in the fifteenth century led to modern science.[4]

Understanding the Myth

To start with a concession, the myth rests on a very modest slice of historical reality, which is then decontextualized and generalized beyond all reasonable boundaries. In fact, early medieval Europe was no hotbed of cutting-edge scientific activity. Neither was the Gobi Desert. Much more important, neither was the city of Rome at the height of the empire. This brings us to two of the major problems with the myth. The first involves confronting the state of scientific knowledge under and after Rome, along with the criteria used to evaluate it. The second involves understanding how the myth developed and the reasons why it has persisted—and continues to be repeated by many people who should know better.

The typical form of the myth is restricted to the European Middle Ages (usually Latin, sometimes Greek). Those who repeat it are interested not in historical understanding but in finding a blunt instrument with which to beat "Christianity" or "the Church" or "Roman Catholicism" or "religion." These entities are invoked to explain the alleged precipitous decline of ancient science, a point that Walker's chart makes nicely.[5] It also illustrates graphically an often-implicit assumption in many discussions of decline, namely that ancient science on its own was on its way and rising (exponentially in the graph). Curiously, things suddenly start picking up—exponentially, once again—in early modern Europe, which was so Christian that people killed one another for their particular flavor of it. Calling the Scientific Revolution "Christian" would, however, irreparably damage the decline-by-Christianity thesis. The adjective is therefore simply omitted.[6]

As its use as a bludgeon leads one to suspect, the myth has little to do with evidence but much more to do with storytelling. Structurally, in its most benign form, the myth of the medieval scientific vacuum fits into a story of revolution. As a matter of narrative, a revolutionary story must depreciate the immediately preceding period, whatever its length. Also as a matter of narrative, it is nonsensical, after alleging a radical break with the past, to discuss that past in careful detail. On the contrary, to be consistent, the revolutionary narrative must undercut the historical connection between the purported revolution and its immediate past. The myth of medieval scientific nothingness is one of the most extensive victims of such a narrative. Adults with a critical sense should, on principle, be skeptical of the claim that in any period of human history nothing happened for a millennium. Curiously, however, they believe it. What's more, the myth-perpetuating disease can infect scholars as well as amateurs. Indeed, this narrative structure is so powerful that even medievalists have been contaminated by it.[7] The narratives of post-1100 medievalists such as myself have also treated the early Middle Ages as a scientific Dark Age, often covering our own ignorance of these early centuries with deprecatory language that valorizes "our" period, from the twelfth to the fifteenth centuries.

No one is arguing that science in the territory that would become medieval Europe stood at a high level in the seventh century. But here is the problem. The alleged precipitous decline of science in the "Christian" Middle Ages is an artifact of a cheap historical trick. Crudely put, it consists of taking the Alexandrian achievements, beginning with Euclid, and spreading them with a trowel over the entire Roman Empire until the murder of Hypatia in the early fifth century. One then goes to the banks of the Seine, Rhine, or Danube and finds a precipitous scientific decline from the cumulative scientific accomplishments of a city without peer in the Roman Empire.

But what standard is this? What, we might ask, was going on scientifically in the third through fifth centuries along the Rhine? Indeed, what was going on in the city of Rome at the height of the republic or empire? To reform the calendar in the first century BCE, Julius Caesar (100–44 BCE) employed Sosigenes of Alexandria: Rome was evidently understaffed.[8] The Roman elites certainly enjoyed nature and could access the outlines of the Hellenistic natural philosophies of the day (Stoic, Epicurean, Neoplatonic, and so on), which were summarized in handbooks and encyclopedias.[9] However, even the educated Romans who read Greek were not interested in the intricacies of Hellenistic mathematical science or natural philosophy or in extending them. If one read only Latin, very few works of Greek science, mathematics, and medicine were accessible.[10] In light of their resources and especially by comparison with the Islamic and Latin civilizations of the Middle Ages, the Romans engaged little with Greek science. Using the mythmakers' criteria, one could easily argue for a precipitous scientific decline in Augustan Rome. In Walker's chart, the rising exponential curve from Greek to Roman culture has the wrong slope.

Where does this state of affairs leave the intellectuals in the Western part of the early Latin Middle Ages? With a serious handicap by comparison with Alexandrian science, which was mostly unavailable. But how could it have been available? Since their Roman colonizers left Greek scientific works largely untranslated, the latter were mostly inaccessible when Latin alone defined literacy. It should be obvious that this situation was well entrenched by the time Constantine legalized Christianity in the fourth century. When Latin Christians expressed lukewarm attitudes toward Greek science, they were reflecting ambient culture, not changing it.

What did happen in the early Latin Middle Ages? Contrary to the myth propagators, the late-antique/early-medieval figure Boethius (ca. 480–524 or 525), a high official from an old Roman family, had planned a large-scale translation of Greek natural philosophical and mathematical works into Latin. His program ended with his execution by the Ostrogothic king Theodoric, a Christian. If Boethius had not been one, too, we could have another Christian barbarian suppressing learning. Too bad history is so complicated.

To gauge attitudes toward science in the Latin Middle Ages, we can conveniently divide the period into two, hinged around the tenth through the eleventh centuries. Institutionally, one of the major differences between the early Middle Ages and the period that immediately follows lies in the locus of control of intellectual training and educational curriculum. Indeed, a change in control often embodies a change in interests.

From the fifth to the eleventh centuries, scholars familiar with Latin—primarily Christian clerics—did their best to collect learned works available in that language, to study them, and, in some cases, to move beyond them. For the reasons previously discussed, the scientific material accessible to them was largely encyclopedic and introductory.[11] There was no translation of Ptolemy's great second-century works in the mathematical sciences. Clearly, the reason why such works were not read in the early Middle Ages is not that "the Church" opposed them (it did not). Rather, as was its prerogative, the civilization of imperial Rome had other priorities than the translation of the Greek scientific enterprise; thus, it exercised its freedom to neglect them. Why, then, should one expect Rome's former colonies in the West to immediately transcend the limitations it had inherited? Eventually they would, but not right away.

Translation as a Symptom of Curiosity

Against this background, we can appreciate two crucial events in the Middle Ages that undergirded the development of early modern science: the translation of Greek manuscripts into Arabic, beginning in the late eighth century, and the later translation of many of these texts into Latin. Each of these translation efforts dwarfed that of the Romans. This double transcultural passing of the scientific baton is a vibrant testimony to the high value placed on Greek science in both Arabic and Latin medieval civilizations. Once translated, Euclid's *Elements* diffused widely in both of them, as did many other fundamental scientific works. Clearly, the myth of medieval scientific nothingness is built on an absurdity. Why would intellectuals in two civilizations waste their lives translating abstruse and complicated works in which they had no interest whatsoever?

The appropriation of Greek scientific learning by Islamic civilization was an unprecedented development in world history.[12] In the eighth and ninth centuries, scholars supported by the caliphs and their intelligentsia sought out and translated Greek and Syriac manuscripts relating to medicine, natural philosophy, astronomy/astrology, mathematics, and the mathematical sciences.[13] From the twelfth through the fourteenth centuries, developments in Islamic science critically and substantively moved far beyond the Greek patrimony it had appropriated.

What is more, word of this success stimulated an analogous translation movement in Latin Europe. From the tenth to the twelfth centuries, dozens of individuals sought out writings of scientific significance and took up the work of translation. Gerard of Cremona (ca. 1114–1187) learned Arabic to translate Ptolemy's *Almagest* into Latin for the first time, plus more than eighty other scientific and medical works from the Arabic, working with Jewish and Muslim associates.[14] This material did not just sit on the shelf unread; it was in demand and fed a hunger for knowledge that access to the Greco-Arabic material promised to satisfy.

Two institutions crucial for science evolved during the medieval period: observatories and universities. Especially in the universities, significant changes in the locus of control and the shape of the curriculum itself helped provide a basic familiarity with scientific knowledge on a vast scale. Both of these developments created a quantitatively significant scientific enterprise with broad social consequences.

Around the later twelfth century, the control of education shifted away from bishops and abbots. In an unprecedented move, masters of arts began to organize themselves into guilds or corporations (the legal term is *universitas*), which secured legal privileges that gave them the autonomy to regulate themselves. Administratively independent of the local bishop and ruler, these new "universities," as they came to be called, determined the require-ments for a degree and the qualifications for becoming a teaching master. The curricula they chose focused overwhelmingly on the newly translated books of Greek, Arabic, and Persian science, in-cluding natural philosophy, medicine, and mathematics. These were works by brilliant pagans and infidels who excited their cu-riosity: Aristotle (384–322 BCE), Euclid (fl. 300 BCE), Claudius Ptolemy (ca. 90–ca. 168), al-Kindi (ca. 801–873), Ibn al-Haytham (ca. 965–ca. 1040), Averroes (1126–1198), and others.

The universities' emphasis on natural philosophy and related disciplines made literacy in these subjects a quantitative phenom-enon with broad socio-intellectual consequences. By one esti-mate, 30 percent of the "arts" curriculum in the new universities focused on knowledge of the natural world.[15] Those who gradu-ated had learned arithmetic, geometry, proportion theory, elemen-tary astronomy, and the new optics, as well as other subjects.[16] Specific universities acquired reputations in particular scientific subjects: Oxford in optics; Oxford and Paris in the science of mo-tion; Bologna, Padua, and Montpellier in medicine; Krakow and Vienna in astronomy.

As the preceding list suggests, universities quickly proliferated, numbering about sixty by 1500. In addition, enrollments in indi-vidual universities grew rapidly. By the middle of the fifteenth century, matriculations had soared to numbers that, in many in-stances, were not surpassed until the late nineteenth and early twentieth centuries.[17] More than a quarter-million students reg-istered in the German universities alone between ca. 1350 and ca. 1500. In short, late medieval Europe had become a scientific culture not only intellectually but also quantitatively. Hundreds of thousands of individuals carried their exposure to Greco-Arabic and indigenous scientific culture into their civic lives—whether or not they graduated. A basic knowledge of natural philosophy and the mathematical sciences was now part of what it meant to be educated, whether the exposure occurred within or outside the dozens of new universities.

Why is this important? In antiquity, the scientific enterprise had been confined to a very small minority of individuals. However large they loom in our own storytelling, they were statistically insignificant (the Museum at Alexandria being one of a kind).[18] The emergence of the universities changed all this by making education in the sciences a permanent and expected part of the intellectual landscape. When Nicolaus Copernicus (1473–1543) came along, he was not a lonely genius born in medieval darkness and fertilized by contact with remote antiquity in Italy. He was one of thousands of university-trained scholars who had not only inherited a widely diffused and reworked amalgam of Greek, Arabic, and Latin scientific learning but also been trained to criticize it (see Myth 3).

THAT BEFORE COLUMBUS, GEOGRAPHERS

AND OTHER EDUCATED PEOPLE THOUGHT

THE EARTH WAS FLAT

Lesley B. Cormack

One of the most enduring myths that children grow up with
is the idea that Columbus was the only one of his time who
believed that the Earth was round; everyone else believed it was
flat. "How brave the sailors of 1492 must have been," you might
imagine, "to travel towards the edge of the world without fear of
falling off!"

—Ethan Siegel, "Who Discovered the Earth Is Round?" (2011)

On the February 15, 2014 Sunday morning program *CBS News
Sunday Morning,* [Charles] Osgood touted a recent speech by
Secretary of State John Kerry in which Kerry "likened deniers of
climate change to those who once believed the Earth is flat."

—Gary DeMar, "Why John Kerry's Flat Earth Society Slam Is
All Wrong" (2014)

Did people in the Middle Ages think that the world was flat? Certainly a quick Google search of the Internet might convince us that this might have been so. Although dozens of Internet sites will tell you that this is a myth to be debunked, the very fact that so many Internet pundits continue to do so tells us just how enduring a misconception this is. The claim that it was once common knowledge that the world was flat is now even a political metaphor for science deniers. As the story goes, people living in the

Dark Ages were so ignorant (or so deceived by Catholic priests; see also Myth 1) that they believed the earth was flat. For a thousand years they lingered in ignorant obscurity, and were it not for the heroic bravery of Christopher Columbus (1451–1506) and other explorers, they might well have continued in this ignorance for even longer. Thus, it was the innovation and courage of investors and explorers, motivated by economic goals and modern curiosity, that finally allowed us to break free from the shackles forged by the medieval Catholic Church.[1]

Where does this story come from? In the nineteenth century, scholars interested in promoting a new scientific and rational view of the world claimed that ancient Greeks and Romans had understood that the world was round, but that medieval churchmen had suppressed this knowledge. Pro-Catholic scholars responded by making the argument that most medieval thinkers readily acknowledged that the world was round.[2] Critics, however, dismissed such opinions as mere apologetics. Why did the battle rage over this particular issue? Because a belief in the flat earth was equated with willful ignorance, while an understanding of the spherical earth was seen as a measure of modernity; the side one defended became a means of condemning or praising medieval churchmen. For natural theologians such as William Whewell (1794–1866) or rationalists such as John Draper (1811–1882), therefore, Catholicism was bad (since it promoted a flat-earth view), while for Roman Catholics, Catholicism was good (since it promoted modernity). As we'll see, neither of these extremes describes the true state of affairs.[3]

This equation of rotundity with modernity also explains why nineteenth-century American historians claimed that it was Columbus and the early mercantilists who proved that the earth was round and thereby ushered in modernity—and America. In fact, it was a biography of Columbus by the American author Washington Irving (1783–1859), the creator of "Rip Van Winkle," that popularized this idea to the world.[4] Although Irving listed a number of arguments made by Columbus, it was his alleged proof of the round earth that struck his audience most forcefully. This equation of a spherical earth with modern scientific thinking also explains why contemporary politicians might want to compare flat-earth thinking with other irrational beliefs, such as denying climate change.

But the reality is more complex than either of these stories. Very few people throughout the Middle Ages believed that the world was flat. Thinkers on both sides of the question were Christians (Roman Catholic or Eastern Orthodox), and for them, the shape of the earth did not equate with progressive or traditionalist views. It is true that most clerics were more concerned with salvation than the shape of the earth—that was their job, after all. But God's works in nature were important to them as well. Columbus could not have proved that the world was round because this fact was already known. Nor was he a rebellious modern; rather, he was a good Catholic who undertook his voyage believing he was doing God's work. A transformation was taking place in fifteenth-century views of the earth, but it had more to do with a new way of mapping than with a move from flat earth to sphere.

Scholars in antiquity developed a very clear spherical model of the earth and the heavens. Every major Greek geographical thinker, including Aristotle (384–322 BCE), Aristarchus of Samos (310–230 BCE), Eratosthenes (273–ca.192 BCE), and Claudius Ptolemy (ca. 90–ca.168), based his geographical and astronomical work on the theory that the earth was a sphere. Likewise, all of the major Roman commentators—including Pliny the Elder (23–79), Pomponius Mela (first century), and Macrobius (fourth century)—agreed that the earth must be round. Their conclusions were in part philosophical—a spherical universe required a sphere in the middle—but were also based on mathematical and astronomical reasoning.[5] Most famous was Aristotle's proof of the sphericity of the earth, an argument used by many thinkers in the Middle Ages and the Renaissance.

If we examine the work of even early-medieval writers, particularly in Europe, we find that with few exceptions they held a spherical-earth theory. Among the early church fathers, Augustine (354–430), Jerome (d. 420), and Ambrose (d. 420) all agreed that the earth was a sphere. Only Lactantius (early fourth century) provided a dissenting opinion, but he rejected all pagan learning because it distracted people from their real work of achieving salvation.[6]

From the seventh to the fourteenth century, every important medieval thinker concerned about the natural world stated more or less explicitly that the world was a globe, and many of them incorporated Ptolemy's astronomy and Aristotle's physics into their work. Thomas Aquinas (1225–1274), for example, followed Aristotle's proof in demonstrating that the changing positions of the constellations as one moved about on the earth's surface indicated the spherical shape of the earth. Roger Bacon (ca. 1214/ 1220–ca. 1294), in his *Opus Maius* (ca. 1270), stated that the world was round, that the southern antipodes were inhabited, and that the sun's passage along the line of the ecliptic affected the climates of different parts of the world. Albertus Magnus (d. 1280) agreed with Bacon's findings, while Michael Scot (1175–1234) "compared the earth, surrounded by water, to the yolk of an egg and the spheres of the universe to the layers of an onion."[7] Perhaps the most influential geographers were Jean de Sacrobosco (1195–1256), whose *De Sphera* (ca. 1230) demonstrated that the earth was a globe, and Pierre d'Ailly (1350–1410), Archbishop of Cambrai, whose *Imago Mundi* (written in 1410) discussed the sphericity of the earth.[8] Both of these books enjoyed great popularity; Sacrobosco's was used as a basic textbook throughout the Middle Ages, while d'Ailly's was read by early explorers, such as Columbus.

The one medieval author whose work has sometimes been interpreted to demonstrate belief in a disk-shaped rather than a spherical earth is Isidore of Seville (570–636), a prolific encyclopedist and natural philosopher. Although he was explicit about the spherical shape of the universe, historians have remained divided on his portrayal of the shape of the earth itself.[9] He claimed that everyone experienced the size and heat of the sun in the same manner, which could be interpreted to mean that sunrise was seen at the same moment by all the earth's inhabitants and that therefore the earth was flat; but the statement more likely implies that the sun's shape did not alter as it progressed around the earth. Much of his physics and astronomy can only be understood to depend on a spherical earth, as does his interpretation of lunar eclipses. While it is not necessary to insist on absolute consistency, it does seem that Isidore's cosmology is consistent only with a spherical earth.[10]

Many popular vernacular writers in the Middle Ages also supported the idea of a round earth. Jean de Mandeville's *Travels to the Holy Land and to the Earthly Paradise Beyond,* written about 1370, was one of the most widely read books in Europe from the fourteenth to the sixteenth centuries. Mandeville explicitly stated that the world was round and navigable:

> And therefore I say sickerly that a man myght go all the world about, both above and beneath, and come again to his own country. . . . And alway he should find men, lands, isles and cities and towns, as are in their countries.[11]

Likewise, Dante Alighieri (1265–1321) in the *Divine Comedy* described the world as a sphere several times, claiming that the Southern Hemisphere was covered with a vast sea. And in "The Franklin's Tale," Geoffrey Chaucer (ca. 1340–1400) spoke of "This wyde world, which that men seye is round."[12]

The one group of medieval writers who explicitly denied the sphericity of the earth came from the Antioch school of theology.[13] Best known was Cosmas Indicopleustes (ca. 550), a sixth-century Byzantine monk who may have been influenced by contemporary Jewish and Eastern flat-earth traditions. Cosmas developed a scripturally based cosmology, with the earth as a tableland, or plateau, placed at the bottom of the universe. It is hard to know how influential he was during his lifetime. Only two copies of his treatise exist today, one of which may have been Cosmas's personal copy, and only one man in the Middle Ages is known to have read his work, Photius of Constantinople (d. 891), widely regarded as the best-read man of his age.[14] In the absence of positive evidence, we cannot use Cosmas to argue that the Christian church suppressed knowledge of the rotundity of the earth. Cosmas's work merely indicates that the early-medieval scholarly climate was open to debates on the subject.

With the exceptions of Lactantius and Cosmas, all major scholars and many vernacular writers interested in the physical shape of the earth, from the fall of Rome to the time of Columbus, articulated the theory that the earth was round. The scholars may have been more concerned with salvation than with geography, and the vernacular writers may have displayed little interest in philosophical questions; but with the exception of the Antioch school in Byzantium, no medieval writers denied that the earth was spherical—and the Roman Catholic Church never took a stand on the issue.

Given this background, it would be silly to argue that Columbus proved the world was round—or even argued so. However, popular accounts continue to circulate the erroneous story that Columbus fought the prejudiced and ignorant scholars and clerics at Salamanca, the home of Spain's leading university, before convincing Queen Isabella (1451–1504) to let him try to prove his position. The group of scholars informally assembled to advise the king and queen of Spain greeted Columbus's proposal— that the distance from Spain west to China was not prohibitively great and that it was shorter and safer than going around Africa— with incredulity. Since no records remain of that meeting, we must rely on secondhand reports written by Columbus's son Fernando (1488–1539) and by Bartolemé de las Casas (1484–1566), a Spanish priest who wrote a history of the New World. Both tell us that the learned men at Salamanca were aware of the current debates about the size of the earth, the likelihood of inhabitants in other parts of the world, and the possibility of sailing through the torrid zone at the equator. They challenged Columbus on his claim to having knowledge superior to that of the ancients and on his ability to do what he proposed. Rather than denying that the earth was spherical, they used its sphericity in their arguments against Columbus, arguing that the round earth was larger than Columbus claimed and that his circumnavigation would take too long to complete.[15]

When Peter Martyr (1457–1526) praised the achievements of Columbus in his laudatory preface to *Decades of the New World* (1511), he was quick to point out that Columbus had proven the equator was passable and that there were indeed peoples and lands in those parts of the globe once thought to have been covered with water. Nowhere, however, did he mention proving the sphericity of the earth.[16] If Columbus had indeed proved the point to doubting scholars, Peter Martyr would surely have mentioned it.

Those who want to preserve Columbus as an icon for the historic moment when the world became round might appeal to the common people. After all, weren't Columbus's sailors afraid of falling off the end of the earth? No, they weren't. According to Columbus's diary, the sailors had two specific complaints. First, they expressed concern that the voyage was taking longer than Columbus had promised. Second, they were worried that because the wind seemed to be blowing consistently due west, they would be unable to make the return voyage eastward.[17]

As we have seen, there is virtually no historical evidence to support the myth of a medieval flat earth. Christian clerics neither suppressed the truth nor stifled debate on this subject. A good son of the church who believed his work was revealing God's plan, Columbus didn't prove that the earth was round—he stumbled on a continent that happened to be in his way.

THAT THE COPERNICAN REVOLUTION
DEMOTED THE STATUS OF THE EARTH

Michael N. Keas

In a wise statement . . . Sigmund Freud argued that all great
scientific revolutions feature two components: an intellectual
reformulation of physical reality and a visceral demotion of
Homo sapiens from arrogant domination atop a presumed
pinnacle to a particular and contingent result, however interesting and unusual, of natural processes. Freud designated two
such revolutions as paramount: the Copernican banishment of
Earth from center to periphery and the Darwinian "relegation" . . .
of our species from God's incarnated image to "descent from an
animal world."

   —Stephen Jay Gould, "Darwin's More Stately Mansion"
   (1999)

Many people believe that Nicolaus Copernicus (1473–1543) de-
moted humans from the privileged "center of the universe" and
thereby challenged religious doctrines about human importance.
This belief is false. Copernicus, a canon in the Catholic Church,
considered his heliocentric (sun-centered) astronomy to be compatible with Christianity, declaring on one occasion that God had
"framed" the cosmos "for our sake."[1] Most other early modern
advocates of heliocentric astronomy similarly affirmed the harmony of the Bible with the new astronomy. However, a few popular authors in the seventeenth century invented an antireligious
story that has become textbook orthodoxy.

The Copernican demotion myth assumes that premodern geo-centricism (earth-centered astronomy) was equivalent to anthro-pocentrism (human-centered ideology). However, according to the ancient Greek geocentric viewpoint that was commonly assumed through the time of Galileo Galilei (1564–1642), earth was at the *bottom* of the universe. This was no honor. "Up" pointed to the exalted incorruptible cosmic heaven; "down" here in the terrestrial realm, things fall apart. The British literary scholar C. S. Lewis summarized the medieval vision of the human place in the cosmos to be "anthropoperipheral."[2] Accordingly, Galileo wrote in 1610: "I will prove that the Earth does have motion . . . and that it is not the sump where the universe's filth and ephemera collect."[3] Galileo offered heliocentrism as a promotion for humanity out of the filthy cosmic center that Dante Alighieri (ca. 1265–1321) had associated with hell.

Johannes Kepler (1571–1630) took this promotional agenda further: We "could not remain at rest in the center" because of "that contemplation for which man was created," which includes acting like "surveyors" who use "triangulation" for "measuring inaccessible objects."[4] Kepler considered moving-earth triangulation methods for calculating planetary distances to be superior to previous geocentric methods. The earth, Kepler argued, was designed for scientific discovery. Although Kepler celebrated the earth's motion, he also (echoing Copernicus) called the sun the "heart of the universe" and a "royal throne."[5] This new "center of importance" language about the sun contributed to the mid-seventeenth-century origin of the Copernican demotion myth.

Kepler—known especially for his "three laws of planetary motion," celebrated in textbooks today—presented a harmonious picture of astronomy and Christianity in his *Epitome of Copernican Astronomy,* the first Copernican textbook. In continuity with earlier challenges to the Aristotelian view that the physical heavens are incorruptible, Kepler's text taught "the truth concerning the mutable nature of the heavens."[6] Kepler alluded to Psalm 102:25–26, in which both heaven and earth are said to "wear out like a garment." Heaven and earth are unified in a finitely durable cosmic dance of eternal significance, he argued.

The Copernican reclassification of the earth as a "planet" spurred more speculation about life on other planets. Kepler thought that extraterrestrial intelligence (ETI) existed, but maintained that it doesn't downgrade human significance. Prior to Kepler, ETI speculation included the medieval idea that God could have made other universes with ETI. Few of these scholastics suggested that God actually made other such worlds, but the possibility of ETI did not count against human dignity.[7] According to a recent survey, most religious people today don't think that the existence of ETI would undermine human significance. "Surprisingly," reports the theologian Ted Peters, "the self-identified nonreligious respondents are the only ones who fear a religious crisis precipitated by contact with extraterrestrials, a crisis expected to happen to others but not to themselves."[8] The ETI "crisis" that secular moderns have invented for their religious contemporaries reinforces the false tale of the Copernican demotion. In fact, ETI speculation in the generation after Kepler contributed to the origin of the Copernican demotion myth.

The Great Copernican Cliché

The scholar Dennis Danielson has described the Copernican demotion myth as the "great Copernican cliché," tracing it to the seventeenth century.[9] An early influential ETI-related example is found in Bernard le Bovier de Fontenelle's (1657–1757) *Entretiens sur la pluralité des mondes* (*Discourse of the Plurality of Worlds*, 1686). The character in the story that speaks for Fontenelle finds himself "extremely pleased" with Copernicus "for having humbled the vanity of mankind, who had usurped the first and best situation in the universe."[10] Danielson explains:

> Once the center was seen as being occupied by the royal Sun, that location *did* appear to be a very special place. Thus we anachronistically read the physical center's post-Copernican excellence back into the pre-Copernican world picture—and so turn it upside down. But

I also suspect (though can't yet prove) that the great Copernican cliché is in some respects more than just an innocent confusion. Rather, it functions as a self-congratulatory story that materialist modernism recites to itself as a means of displacing its own hubris onto what it likes to call the "Dark Ages." When Fontenelle and his successors tell the tale, it is clear that they are making no disinterested point; they make no secret of the fact that they are "extremely pleased" with the demotion they read into the accomplishment of Copernicus.[11]

Although the demotion myth originated in the mid-seventeenth century, it did not enter English-language astronomy textbooks until the nineteenth century, when it became the centerpiece of a popular textbook story of scientific triumph over religious superstition.

In his *Treatise on Astronomy*, Horatio N. Robinson (1806–1867), a sometime mathematics teacher in the United States Navy, depicted early Copernican advocates as antireligious warriors:

> The true solar system . . . is called the Copernican system. . . . But this theory, simple and rational as it now appears, and capable of solving every difficulty, was not immediately adopted; for men had always regarded the Earth as the chief object in God's creation; and consequently man, the lord of creation, a most important being. But when the Earth was hurled from its imaginary, dignified position, to a more humble place, it was feared that the dignity and vain pride of man must fall with it; and it is probable that this was the root of the opposition to the theory.[12]

Contrary to Robinson, early modern astronomers resisted a moving earth primarily for scientific, not theological, reasons. During Galileo's career, Tycho Brahe's (1546–1601) geoheliocentric system was widely believed to best fit the prevailing physical theory and observations, such as the phases of Venus, which show it circles the sun. In the Tychonic system, the sun and fixed stars revolve around a central-stationary earth. The planets revolve around the sun as their (moving) center. Although Isaac Newton's (1643–1727) physics displaced Tycho's cosmology, many Copernican problems remained unsolved until about the time of Robinson's 1849 textbook.[13]

Robert Cardinal Bellarmine's (1542–1621) pivotal April 1615 letter aimed at Galileo is consistent with the historical assessment that theology did not provide the main rationale for the early modern anti-Copernican scientific consensus. This leading Catholic theologian of the Inquisition explained that if Copernicanism were to be firmly established by a "true demonstration" (which he considered extremely unlikely), "then one would have to proceed with great care in explaining the Scriptures that appear contrary."[14] But Galileo failed truly to demonstrate the Copernican system. Immature science, not restrictive religion, was to blame. Robinson's 1849 tale of a Copernican challenge to human dignity is nowhere in Bellarmine's letter. Indeed, such concerns had very little to do with this otherwise embarrassing episode in Catholic history.

The Copernican Demotion Myth Grows

A century after the publication of Robinson's textbook, Cecilia Payne-Gaposchkin's (1900–1979) *Introduction to Astronomy* (1954) expanded the Copernican demotion to include later discoveries. "The advance of astronomical knowledge," she declared, "has successively dethroned the Earth, the Sun, and the stellar system from their supposed unique and central stations."[15] Two years earlier, in his book *Cosmology,* Hermann Bondi (1919–2005) had coined the term "Copernican principle" to refer to this extended Copernican demotion story.[16] We shall see how Bondi's terminological invention breathed new life into the demotion narrative.

Most astronomy textbooks today perpetuate the extended demotion myth. *The Cosmic Perspective* (2014), a widely used introduction to astronomy, features a foreword by Hayden Planetarium director Neil deGrasse Tyson, who frames the Copernican achievement as the first of multiple humiliating discoveries. Ironically, humans are still special, Tyson suggests. How so?

Because "the cosmic perspective is spiritual—even redemptive—but not religious." Tyson believes that the Copernican demotion story is redemptive because it saves us from ignorance and offers secular spirituality. He recommends the spiritual feelings he has experienced while watching shows at his own Hayden Planetarium:

> I feel alive and spirited and connected. I also feel large, knowing that the goings-on within the three-pound human brain are what enabled us to figure out our place in the universe.[17]

Sociologist Elaine Howard Ecklund recently found that such secular spirituality is common among scientists. Among spiritually engaged scientists, 22 percent self-identified as atheists, 27 percent as agnostic, and the rest (51 percent) as theists. The nontheistic "spiritual" scientists reported experiences of awe in the face of nature.[18] Such ennobling naturalistic spirituality is present in many astronomy textbooks today, and ironically, it is often connected to the standard dethronement narrative.[19]

Dennis Danielson analyzes this spirituality as it is justified by an alleged series of human dethronements beginning with Copernicus:

> But the trick of this supposed dethronement is that, while purportedly rendering "Man" less cosmically and metaphysically important, it actually enthrones us modern "scientific" humans in all our enlightened superiority. It declares, in effect, "*We're truly very special because we've shown that we're not so special.*" By equating anthropocentrism with the now unarguably disreputable belief in geocentrism, such modern ideology manages to treat as nugatory or naive the legitimate and burning question of whether Earth or Earth's inhabitants may indeed be cosmically special. Instead it offers—if anything at all—a specialness that is cast in exclusively existential or Promethean terms, with humankind lifting itself up by its own bootstraps and heroically, though in the end pointlessly, defying the universal silence.[20]

Danielson's commentary captures the pervasive dethronement—but *subtle enthronement*—narrative in astronomy textbooks since the nineteenth century. Consider Eric Chaisson and Steve Mc-Millan's introductory astronomy textbook in light of Danielson's analysis:

> Yet there was a time . . . when our ancestors maintained that Earth had a special role in the cosmos and lay at the center of all things. Our view of the universe—and of ourselves—has undergone a radical transformation since those early days. Humankind has been torn from its throne at the center of the cosmos and relegated to an unremarkable position on the periphery of the Milky Way Galaxy. But in return we have gained a wealth of scientific knowledge. The story of how this came about is the story of the rise of science.[21]

The expression "unremarkable position on the periphery of the Milky Way Galaxy" is worth pondering. It shows how the extended Copernican demotion myth often influences scientific assessment of the earth's significance today. Since the astronomers Guillermo Gonzalez and Donald Brownlee introduced the "galactic habitable zone" (GHZ) in a 2001 paper, studies have largely confirmed this restricted life-friendly zone within a galaxy.[22] Most astronomy textbooks now discuss the GHZ and acknowledge that a planet near a galaxy's center (or too distant from the galactic center) would be uninhabitable owing to multiple factors. Curiously, most textbooks, including the one just mentioned, still depict our knowledge of the earth's noncentral location within our galaxy as a humiliating demotion. The demotion narrative diminishes the significance of our GHZ location without appeal to evidence.

The humiliation interpretation of our noncentral place within the Milky Way galaxy traces back to Harlow Shapley (1885–1972)—the man who discovered the extent of the galaxy in 1918. Most astronomy textbooks still celebrate Shapley's discovery *and* perpetuate the metaphysical meaning that he imposed on it. This astronomy education tradition goes back to at least Payne-Gaposchkin's 1954 textbook, which we previously inspected.

This should be no surprise as, indeed, Shapley was Payne-Gaposchkin's PhD adviser and Harvard colleague. As America's celebrity astronomer prior to Carl Sagan, Shapley had launched a "lifetime assault on anthropocentric thinking"—so a leading Shapley scholar reports.[23]

Evaluation of human significance in the light of science has taken several new turns since Shapley's crusade against anthropocentrism. A recent NASA publication says it all. After essays by leading astrobiology scholars such as Eric Chaisson and Steven Dick, NASA scientist Mark Lupisella (who coedited this book with Dick) writes the following:

> *A Cosmic Promotion?* Scientists and thinkers have been fond of pointing out humanity's "great demotions." From Copernicus to modern day cosmology (perhaps with the exception of "anthropic principles" and associated observations of "fine tuning"), humanity has been displaced and demoted from privileged positions in the cosmos. Perhaps it's time for a promotion—one that goes beyond the confusion of anthropic principles, one that does not rely on teleological assumptions and assertions about the ultimate nature of the universe. Bootstrapped cosmocultural evolution allows for the possibility that life, intelligence, and culture could have arisen by chance, while at the same time asserting that such phenomena are cosmically significant. Stronger versions suggest that cultural evolution may have unlimited significance for the cosmos. Our cosmic location and means of origin should not be confused with our cosmic potential.[24]

Will this enthusiastic vision of "our cosmic potential" overwhelm the dethronement narrative that has dominated astronomy textbooks since the nineteenth century? Jeffrey O. Bennett's *The Cosmic Perspective* (2014) has already embraced Tyson's Sagan-like mantra: "The cosmic perspective is spiritual—even redemptive—but not religious."[25] Compared to the medieval, Copernican, and Keplerian accounts of cosmic human dignity rooted in Christianity, such twenty-first-century proclamations of the "redemptive" effect or "unlimited significance" of cultural evolution have abandoned cosmological modesty.

The Great Copernican Equivocation

How did the Copernican demotion myth become the dominant textbook narrative? Chaisson and McMillan's *Astronomy Today* (2014) offers a clue. After surveying evidence for the Copernican system, the authors state the obvious: "Today, the evidence is overwhelming." But then they go on to claim, "This removal of the Earth from any position of great cosmological significance is generally known, even today, as the *Copernican principle*. It has become a cornerstone of modern astrophysics."[26]

Unlike other science–religion conflict myths, the Copernican demotion has been canonized as an astronomical "principle" that bears the name of a scientific saint. The rhetorical strategy is to equivocate between the now unassailable argument for a moving earth and the contested case for our cosmic insignificance. In the name of Copernicus, the evidence for both is declared "overwhelming." Most readers are no doubt unaware of the equivocation in operation here. As we evaluate the merits of the Copernican principle today, let us remember that Copernicus himself—indeed, most early-modern astronomers—did not actually embrace the idea.

THAT ALCHEMY AND ASTROLOGY WERE SUPERSTITIOUS PURSUITS THAT DID NOT CONTRIBUTE TO SCIENCE AND SCIENTIFIC UNDERSTANDING

Lawrence M. Principe

Astrologic nonsense . . . has done so much harm in the Western world and is still poisoning weak-minded people today.

— George Sarton, *A History of Science* (1952)

The alchemists began with self-deception and rather than admit their failure and foolishness they piled up new fantasies upon the old ones; they were fools or knaves or more often a combination of both in various proportions.

— George Sarton, "Boyle and Bayle: The Sceptical Chymist and the Sceptical Historian" (1950)

As the epigraphs by the influential historian of science George Sarton (1884–1956) suggest, astrology and alchemy are widely considered to have been foolish and irrational pursuits. Many people think of them as "prescientific" follies that retarded the development of the modern sciences of astronomy and chemistry. Such ideas are wrong, arising from a mistaken understanding of both the content and the practice of alchemy and astrology in their heydays prior to the eighteenth century. In fact, both astrology and alchemy contributed substantially and positively to the development of modern science, even if some of their starting assumptions may since have been proven to be incorrect.[1]

Serious astrology should never be confused with modern-day popular horoscopes that purport to tell your fortune based on your zodiacal sign. That trivialized form of astrology claims, more often for entertainment than for anything else, to predict a person's fate or the events of one's day based solely on when in the year he or she was born—that is, where the Sun was in its annual path through the zodiac. The only (and scant) commonality that this form of astrology shares with premodern astrology is the notion that celestial bodies exert some kind of influence on the earth. But serious premodern astrology considered the positions of all heavenly bodies (not just the sun) at a precise moment (not an entire month) to calculate a complex combination of possible effects. Moreover, most astrologers proposed naturalistic (*not* magical) mechanisms for the operation of celestial effects on earth and on the human body.

Although the exact mechanism and extent of action remained debated, there were no doubts about the reality of celestial effects on the earth. Premoderns observed how the ocean rises and falls with the position of the Moon, and how women's bodies also obey lunar phases—the menstrual cycle averages twenty-eight days, exactly the period of the Moon's orbit around the earth. Hotter weather occurs when the sun reaches a certain region in the heavens, and cold returns after it leaves that region. A magnetized needle always turns spontaneously to point toward the North Star. How could these natural phenomena occur unless invisible celestial influences acted on terrestrial things? Thus, observations of the natural world, not superstition, lay at the foundation of premodern astrology.

Astrology endeavored to gain useful knowledge and to prognosticate about the weather, natural and human events, and especially human health.[2] In medicine, prevailing celestial influences were thought to "imprint" a newborn at the moment of birth with a unique set of bodily properties (called a "complexion").

Different complexions led to different strengths and weaknesses, and greater resistance or susceptibility to particular diseases and personality traits. For example, a child imprinted strongly by a "cold, dry" planet such as Saturn might tend toward melancholy or sluggishness (what we call depression). Armed with such diagnoses, people could then adapt diet, lifestyle, and behavior to prevent the problems to which they were prone and be alert to upcoming celestial alignments that might trigger certain bodily (and thence behavioral) responses. The goal was greater control over one's life and health through a fuller knowledge of one's physical and mental makeup and of the external influences on it. Modern physicians hope to do much the same with genetic analysis.

Critical to advanced astrology was the conviction that celestial bodies "incline but do not compel" and that the "wise person will master the stars." These maxims mean that astrological influences neither control us nor cause events to happen. Instead, by knowing astrological effects and exercising free will, we can prevent illnesses as well as avoid unwise decisions when clear thinking might be impaired by such influences.[3]

To gain such knowledge, the positions of all celestial bodies relative to the horizon had to be calculated precisely at exact times. This calculation was known as casting a *horoscope*—literally, an "observation of the time"—and required substantial mathematical and technical skills. Thus, serious astrologers had to be intelligent, skilled, and observant people. The pursuit of astrology led to the development of new mathematical tools to make calculations easier—for example, in spherical trigonometry. In fact, in the premodern period, the title "mathematician" often actually referred to an astrologer. Astrology also required improved celestial observations to identify astronomical cycles and their exact periods, as well as the creation of new astronomical models and tables. If exact planetary positions could not be calculated accurately many years into the past or the future, then there was great danger of misreading a person's complexion or the prevailing influences at a given time and thus giving the wrong advice. Planetary motions had to be observed and charted with enormous accuracy, and astronomical models were constantly refined in order to account for observations and to better predict future motions. Even such fine variables as how much the earth's atmosphere refracted incoming light had to be taken into consideration. Hence, the demand for reliable and precise astrological data frequently drove innovation, observation, and discovery in astronomy.

Many of the greatest names in astronomy were also involved in (or even inspired by) astrology. Claudius Ptolemy (ca. 90–ca. 168) is famed for his *Almagest,* a summary of ancient Greek mathematical astronomy, which provided the fundamental view of the cosmos for fifteen hundred years. But he also wrote the *Tetrabiblos,* a survey and exploration of how to use astronomical knowledge for astrology. Tycho Brahe (1546–1601), the greatest and most prolific naked-eye astronomer, worked in astrology and took up astronomical observation in large part to improve the accuracy of astrological tables. Even Galileo Galilei (1564–1642) cast birth horoscopes for patrons as well as for himself and his children. Robert Boyle (1627–1691) tried to capture celestial influences—thinking they consisted of material particles—in chemical substances, and pondered the action of celestial bodies on the salubrity of the air.[4]

Failure to provide accurate advice and predictions generally undercut the reputation of individual astrologers but not of astrology itself. Celestial positions could be calculated unambiguously, but evaluating the cumulative effect of seven planets radiating counteracting influences from twelve distinct regions of the sky was difficult and ambiguous. Modern economics provides a good comparison. Economists, who are often called on to prognosticate about future economic trends, are rarely (if ever) unanimous in their predictions, in part because, like astrology, economics deals with highly complex interacting systems. Yet such disagreements and frequent failed predictions do not make us abandon economics; instead, they leave us hoping for better future understanding and accuracy.

Alchemy often suffers from an equally bad reputation among the general public, often being derided as "magic" or sorcery, a foolish fancy, or simply a fraud. But such judgments ignore historical facts. The English Franciscan friar Roger Bacon (ca. 1212/1220–ca. 1294) clearly defined alchemy as consisting of theories that "speculate about all inanimate things and the entire generation of things from the elements" and practices that "teach how to make precious metals, colors, and many other things better and in greater quantity than is done by nature."[5] This definition and many others like it reveal alchemy as a pursuit very much akin to chemistry. Indeed, the words "alchemy" and "chemistry" were interchangeable until at least 1700.[6] Alchemy made use of both head and hand, theory and practice. Throughout its history (from the second to the eighteenth century), it remained a practical endeavor that studied the properties and transformations of material substances, aiming to use such knowledge toward productive ends.[7]

Alchemy is most frequently connected with the aim of making gold from cheaper metals, a process known as transmutation. Although modern chemistry now denies the possibility of this transformation, the alchemists based their hope on what were at the time solid theoretical and observational grounds. Modern chemistry sees the metals as non-interconvertible *elements,* but alchemists thought the metals were *compounds,* arising from the subterranean combination of two (or more) simpler substances. The seven metals known to them were produced from the combination of these simpler substances in differing proportions or grades of purity. For example, lead and tin had too much of the "liquid" principle (often called Mercury), as shown by their low melting points. Copper and iron had too much of the "dry" or inflammable principle (often called Sulphur), hence their difficult fusion and ability to burn. Thus, it should be possible to adjust the relative proportions of these ingredients and so transform one metal into another. This transmutation appeared to happen naturally, but very slowly, in the earth. The alchemists observed (correctly) that silver ores generally contain some gold, and lead ores some silver, as if the metals were slowly being purified and transformed underground into better ones. The alchemists labored to find the means of carrying out the process more quickly and efficiently.[8] They endeavored to prepare a variety of chemicals that would bring about the transformation; the most powerful and sought after of these substances they called the Philosophers' Stone or Elixir.

But the quest for transmutation and the Stone was only one of many alchemical pursuits. Practitioners also sought to prepare better medicines, often by chemically extracting, treating, or purifying natural substances such as those found in plants. They made new pigments, dyes, cosmetics, salts, glass, distilled liquors, and metallic alloys. They looked for ways to imitate nature and to improve on naturally occurring materials. They contributed to better methods for working and smelting ores.[9] In so doing, they discovered, prepared, and described new substances and developed many of the methods for manipulating chemical substances still used routinely by chemists today—distillation, sublimation, crystallization, and so on—as well as techniques for assaying and analysis.

Alchemists' contributions were not solely in the area of material production and practical methods. The alchemists also theorized about the hidden nature of matter and its composition, and used such theories to explain and to guide their experimentation. By the late Middle Ages, some had begun to develop quasiparticulate matter theories—that is, the idea that material substances were composed of more or less permanent but invisibly small particles. Such ideas eventually fed the revival of atomism in the seventeenth century, and alchemical experiments provided the best available evidence for such theories. Some scholars have pointed to alchemy as a major source for the routine use of experiments to learn about the natural world—a key feature of modern science—and cite alchemy as the earliest and most compelling advocate for the ability of human beings to change the world through technology.[10]

There can be no question that the accumulated knowledge and experience of the alchemists formed the foundations on which modern chemistry was built. The serious interest and fervent activity in alchemy by important figures of the Scientific Revolution is now well known. Robert Boyle, often called the father of modern chemistry, spent forty years trying to make the Philosophers' Stone. He claimed to have witnessed transmutations at the hands of others and, in 1689, convinced Parliament to strike down an old law forbidding transmutation. He drew on alchemical theories and practices for developing his own ideas, and received his first serious training in laboratory operations from one of the most celebrated alchemists of the seventeenth century, the Harvard-trained George Starkey (1628–1665), aka Eirenaeus Philalethes. Isaac Newton's (1643–1727) long-term pursuit of alchemy is now well recognized. He spent years copying and comparing texts to uncover alchemical secrets, and maintained a laboratory in which he spent countless hours on alchemical experiments. In 1692, he had the philosopher John Locke (1632–1704), who was also interested in alchemy, rifle through the papers of the recently deceased Boyle in search of alchemical recipes to send him.[11]

The point to remember is that while moderns may link both astrology and alchemy to superstition or "magic," the actual practitioners of the past saw their fields as fully naturalistic. Connections to the occult were forged only in the eighteenth and, with more frequency, the nineteenth centuries.[12] While there always exists a range of ability and intelligence within any group of practitioners, it remains clear that serious astrologers and alchemists were sober explorers of nature no less than good modern scientists. The bad reputations of these fields of inquiry arose in large part in the eighteenth century from those who wanted to aggrandize their own originality and importance by dismissing the work and achievements of their predecessors, publicizing abuses and errors by the worst practitioners rather than acknowledging the achievements of the best. Nevertheless, modern historical research continues to demonstrate the true character of astrology and alchemy and has firmly reinstalled them as important contributors within the history of science.

THAT GALILEO PUBLICLY REFUTED ARISTOTLE'S CONCLUSIONS ABOUT MOTION BY REPEATED EXPERIMENTS MADE FROM THE CAMPANILE OF PISA

John L. Heilbron

Many of Aristotle's conclusions about motion were held to be
very clear and indisputable. . . . [Galileo showed the contrary] by
repeated experiments made from the height of the Campanile of
Pisa in the presence of the other teachers and philosophers and
all the students.

—Vincenzio Viviani, *Vita di Galileo* (1654)

The incident recorded in the epigraph refers to a demonstration
that Galileo Galilei (1564–1642) may or may not have conducted
when he was a young lecturer at the University of Pisa. The clear
and indisputable conclusions he thus publicly sustained against the
Master of Those who Know were that "the velocities of bodies of
different weights made of the same material moving through the
same medium do not keep the proportion of their weights . . . but
they are all moved with equal velocities," and that "the velocities of
the same body moving through different media retain the reciprocal proportion of the resistances or densities of these media." We
know this incident from Vincenzio Viviani (1622–1703), who
probably learned its ingredients from Galileo, with whom he lived
as pupil and assistant during the last years of Galileo's life. The

legend first appeared in the short biography that Viviani wrote as front matter to a projected edition of Galileo's writings, aborted under pressure from the Roman Catholic Church.[1]

There are several reasons for suspecting that Viviani's story is a myth: No member of the large and literate audience supposedly present and shocked by Galileo's disproof of received truth seems to have written a word about it. The theses ascribed to Aristotle (384–322 BCE) are taken out of context. Galileo would have been as surprised as his audience had his repeated experiments showed that all bodies fall with the same velocity. Nonetheless, much of Viviani's story can be confirmed if we take it to apply to a group rather than to an individual. It is true and false in ways that lend themselves to pedagogy and myth. Myth mongers who have added such details as the time of day, the suspense of the expectant viewers, and the weights and materials used have made the legend the exemplar of the triumph of commonsense experiment over slavish adherence to authority—and also, perhaps, of storytelling in the history of science.[2]

The reason that none of the putative witnesses to Galileo's singular performance from the campanile ever mentioned it is that it did not take place. In any case, the demonstration would not have been singular. Professors claiming to know and show the truth about motion, including Galileo's professors when he studied at Pisa, Girolamo Borro (1512–1592) and Francesco Buonamici (1533–1603), frequently threw objects from the windows of their lecture rooms. Borro reported that as often as he dropped chunks of wood and lead balls of roughly the same weight, the wood at first fell faster than the metal, which, he said, answered the tricky question whether air has weight (tends to descend) in air. He apparently had a talent for observation. Filmed re-creations of his demonstration confirm his results, which Galileo also accepted when he taught at Pisa. The effect apparently arises from the experimenter's unconscious tendency to release the wooden object first.[3]

During his tenure at Pisa, from 1589 to 1592, Galileo probably added his heavy missiles to the dangers of studying philosophy there. In a manuscript "De Motu," unpublished in its time but known to Viviani, he refers to Borro's experiments and adds that the lead soon overtakes the wood, "and if they are let fall from a high tower, precedes it by a large distance, and this I have often tried by experiment."[4] If Galileo did these experiments, his students probably attended them, just as he had witnessed Borro's and Buonamici's; and if the tower was the cathedral's campanile, the choice was so obvious that Giorgio Coresio, one of Galileo's philosophical opponents, "[whose] intellect could no more understand Aristotle than an anvil can fly," used it to try to confirm the standard account of motion.[5]

The passages in Aristotle to which Viviani referred and in which later mythographers have spied the hidden algebra "v[elocity] = W[eight]/R[esistance]" do not concern free fall directly. Rather, they provide the hinges for an argument against the existence of a vacuum understood as a space with no properties but extension. What would happen to a moving body entering such a space? Well, it would have no idea where or what it was, no notion of up or down, no way of knowing whether it was a rock or a balloon. This is because the locomotion of a body—or, if at rest, its tendency to motion—depends on its place, which Aristotle defined, after much fuss, as the interior surface of the containing medium.[6] Water in air proceeds, or has a tendency to proceed, downward because it is in air; if in sand, it would want to move upward; if in a vacuum, it would not move at all, a void "having no place to which things can move more or less than to another." Or, rather, if set in motion, it would move ad infinitum, "for why should it stop *here* rather than *here?*"[7]

As further proof, Aristotle adduced this hostage to fortune: "we can see the same weight or body moving faster than another for two reasons," either because of the nature of the medium or because of an "excess of weight or lightness." To this he added a compromising illustration: if air is twice as "thin" as water, a body heavier than water will traverse a given distance in air twice as fast as in water. "And always, by so much as the medium is more incorporeal and less resistant and more easily divided, the faster will be the movement." Hence a body, any body, would have an infinite velocity in vacuum, which is absurd.[8] Again, since observation shows that "bodies that have a greater impulse either of weight or of lightness, if they are alike in other respects, move faster over an equal space, and in the ratio that their magnitudes bear to one another." This is because the greater the weight, the more easily it divides the medium it penetrates. If there is nothing to divide, all bodies would move with the same speed. "But this is impossible."[9] Thus, for two good and sufficient reasons there can be no void, and therefore—here is the point of the business— we must reject the theories of the atomists.

It is perverse to read these passages as expressions of quantitative relations intended seriously. All Aristotle wanted from his inverse proportion of speed to resistance was to put "thinness" in the denominator, and his implausible numerical example, assuming air to be twice as thin as water, could not have been meant literally. Similarly, the vague statement that bodies with "a greater impulse . . . of weight" move faster in proportion to their magnitudes when traversing a plenum was made not to lay down the law $v \propto W$ but to secure what Aristotle regarded as a reductio ad absurdum: since there is no function for weight in a vacuum, all bodies would move with the same velocity there.

It is also perverse to read the discussion, which pertains to directionless motion in a void, as an account of falling bodies. But that is precisely what Galileo did when "quoting" Aristotle to the strawman peripatetic philosopher who is the butt of the dialogue in *Two New Sciences* (1638), in which Galileo first made public his definitive theories on motion. After Galileo's spokesman admits that occasionally he exaggerates a little, he proceeds as follows:

> Aristotle says, "A ball of iron of a hundred pounds falling from a height of a hundred *braccia* hits the ground before one weighing one pound descends a single *braccio.*" I say that they arrive at the same time; you [may] find when making the experiment that the larger precedes the

smaller by two fingers' breadth . . . will you hide Aristotle's ninety-nine *braccia* behind these two fingers?[10]

Two fingers to Aristotle! How could he be so ignorant about the most obvious things?

Galileo had expressed himself with equal confidence about these obvious things in 1590, although the theory of motion he held then predicted experimental results significantly different from those claimed in *Two New Sciences*. The theory of 1590 or thereabouts anticipated that in free fall, light objects at first descend more swiftly than heavy ones, and that if the drop is high enough, a body will attain a velocity proportional to the difference between its specific gravity and that of air. "Oh subtle invention, most beautiful thought! Let all philosophers be silent who think they can philosophize without knowledge of divine mathematics!" What about the corresponding Aristotelian teachings about motion? "Oh ridiculous chimeras! Immortal gods, how, please, can anybody believe in them since the contrary is obvious to sense?"[11]

The longest convenient vertical drop from the Leaning Tower is about 150 feet. Ignoring air resistance, Galileo's weights would have taken just over three seconds for their fall. He and his students would have had trouble discerning by eye or ear whether the weights reached the ground together ("within two fingers' breadth") or not. Hence, they could not have confirmed the law that Galileo then did not hold: that the speed of fall (absent resistance) is the same for all bodies. Nor could they disconfirm by his experiments that a body passes through different media with velocities in "reciprocal proportion to [their] resistance or densities," since they examined only descent in air. However, they could have confirmed that the velocities of different bodies do not "keep the proportion of their weights," since if they did, a body twice the weight of another would precede it by a second or so.

In any case, it is hard to drop weights by hand at the same instant precisely along the vertical. When Galileo's remote successor at Pisa, Vincenzo Renieri (1606–1647), tried the experiment from the campanile in 1641, he could not confirm the results Galileo gave in his definitive treatment of the problem in *Two New Sciences*. The circumstances of Renieri's failure very probably played a part in the creation of the legend. Being informed (so he wrote Galileo) that "some Jesuit" (the adept natural philosopher Niccolò Cabeo [1586–1650]) had found that unequal weights fell to the ground from the same height in the same time (in agreement with Galileo's definitive law) and *doubting the result,* Renieri tried the experiment and saw the lead ball hit the ground four or five feet before the wood. Repeating it with lead balls of different weights, he confirmed that the heavier landed before the lighter. Also, it seemed to him that toward the end of its fall, a wooden object tended to swerve.

A lost letter from Galileo must have reminded Renieri that the definitive law of free fall could be found in *Two New Sciences*. Renieri's reply to the reminder is almost as hard to credit as the legend itself. He had not read Galileo's long-awaited book! He excused himself by blaming the heavy teaching load that had kept him from it for over two years. If true, it is instructive. Did he think *Two New Sciences* too demanding to be read during the school year and too insignificant to have priority during vacations? It is more likely that he knew what he took to be Galileo's views from conversations at Galileo's home prison in Arcetri, which he visited frequently as a friend, consoler, and collaborator. He had retained from these conversations that different bodies do not fall with velocities proportional to their weights; Cabeo's assertion that lead and a crust of bread fell together seemed to him too extravagant an application of the principle. Renieri had also retained from Galileo, "what I thought I heard or read from you," that the larger of two weights of the same material falls faster. He closed his letter reporting his heresies by sending his most affectionate greetings to Viviani.[12] Since by 1641 Galileo was totally blind, Viviani would have read him Renieri's letters and helped with the replies. No doubt the process prompted some conversation about fall experiments at the Leaning Tower between the master and his assistant. As every oral historian knows, the dialogue would have proceeded along these lines:

> *Viviani:* What do you say about Sig. Renieri's experiment, maestro?
>
> *Galileo:* It is not easy. I tried something of the same kind about fifty years ago, but then I knew the outcome I expected.
>
> *V:* Did things work out as you expected?
>
> *G:* Pretty closely. I tried the experiment several times with several people present—my students and a few other professors and philosophers. Old Buonamici came once, and also Jacopo Mazzoni, who brought some of his many students. He was working then on his famous comparison of the philosophies of Plato and Aristotle and looking for mistakes Aristotle made by neglecting mathematics.
>
> *V:* And of course Aristotle would have seen the inanity of his doctrine of motion if he had followed out the logic of the proportions he set down, as you, maestro *mio*, have demonstrated so clearly.
>
> *G:* No doubt. . . . Hmmm, Yes, Mazzoni must have been there. He was a family friend, and I talked with him a lot about the big problems in philosophy and helped him develop examples against Aristotle's nonmathematical way.[13]
>
> *V:* So fifty years ago you did experiments like Renieri's from the campanile and showed them to a few teachers and students? And the results made clear to them that Aristotle's pseudo-mathematics of motion was nonsense? And that all bodies fall at the same speed apart from the slowing effected by the medium?
>
> *G:* That's about right. But of course it was a long time ago, and it took me many years to perfect my ideas about motion as you find them in *Two New Sciences.*

In reworking his materials, Viviani combined a faithful Pisan portrait (the hail of falling bodies, the audience of students and professors, the theatrical test of Aristotle) with artistic touches that gave life to the legend (Galileo as sole performer, the campanile as test site, the test as a choice between a caricature of Aristotle's views and a misrepresentation of Galileo's). The maestro's writings abound in such touches. He portrayed himself as the victim when he was the aggressor. He created and annihilated straw-men opponents. And he exaggerated the accuracy and reliability of his results to the limits of his powerful rhetoric.[14] His caricature of Aristotle's "law of falling bodies" is grosser by far than Viviani's. Galileo told stories about the physics he invented; Viviani told lesser fibs about the manner in which the maestro's inventions took place. The legend is an excellent piece of seventeenth-century rhetoric contrived jointly by teacher and disciple to promote themselves and the application of mathematics to physics, and to push back the date of Galileo's definitive teaching about motion to the time of his first experiments on the subject. As the great Eusebius (ca. 275–339) understood when incorporating the Old Testament in his history of the Christian church, an orthodoxy is the more persuasive the longer its pedigree.

It appears that exploration of the origins of the legend of the Leaning Tower may help the historian identify and appreciate some of the differences in the practice of physics between Galileo's day and ours. Further exploration will suggest other applications. An analysis of the practical difficulties of the tower experiment and of the finer details of free fall as described in Newtonian mechanics could make good pedagogy. The reasons for propagating and embellishing the legend might also repay study. Of these, the least interesting is the celebration of the triumph of one notion of motion over another. More significantly, the legend conveys answers to such fundamental questions as whether physics should be qualitative, coherent, and explanatory, or quantitative, piecemeal, and descriptive. Perhaps of most importance in the present context, the myth mongers who purvey the legend without looking into it commit the very crime that it warns against: they rely on authority, on someone else's say-so, rather than on their own informed judgment.

THAT THE APPLE FELL AND NEWTON INVENTED THE LAW OF GRAVITY, THUS REMOVING GOD FROM THE COSMOS

Patricia Fara

A different challenge to religion arose with Isaac Newton. His theories of motion and gravitation showed how natural phenomena could be explained without divine intervention.

—Steven Weinberg, "On God, Christianity and Islam" (2007)

Quite simply, Darwin and Wallace destroyed the strongest evidence left in the 19th century for the existence of a deity. Two centuries earlier, Newton had banished God from the clockwork heavens. Darwin and Wallace made the deity equally redundant on the surface of the earth.

—Johnjoe McFadden, "Survival of the Wisest," *The Guardian* (2008)

A 1970s advertising caption for London's *Financial Times* acclaimed Isaac Newton (1643–1727) as "the British physicist linked forever in the schoolboy mind with an apple that fell and bore fruit throughout physics."[1] Even schoolgirls are familiar with Newton's apple, but many scientists also know—or at least think they know—that Newton eliminated God from the cosmos. These are two different types of story. Historians can never be sure whether or not Newton was indeed inspired by an apple; in contrast, they are absolutely certain that Newton was a deeply religious man who believed that God is ever present in the universe.

The factual truth of the falling apple is not particularly important: what matters is its symbolic significance as the founding moment of Newtonian physics. It resembles other romanticized episodes of dramatic discovery, such as Archimedes's (ca. 287–ca. 212 BCE) shout of "Eureka" from his bath, or James Watt's (1736–1819) childhood fascination with a boiling kettle. In these eureka versions of history, theories are born fully fledged in the mind of a scientific genius, in the same way that a symphony or a poem might inexplicably arise in the brain of a musician or writer. The quasi-historical details convert famous real-life heroes into mythological characters who influence how people think about science.

The claim that Newton banished God from science is not a scientific myth; it is a wrong belief that can be firmly rejected on the evidence of Newton's own published writings. Those who promulgate the no-God fallacy fail to appreciate that Newton's original version of Newtonian physics is not the same as the one that prevails now. Newton's God did not create the universe and then retire; instead, he permeated his creation and intervened from time to time in its operation—which is why critical contemporaries accused Newton of supposing God to be a sloppy clockmaker.

The Apple Myth

Toward the end of his life, Newton told the apple anecdote around four times, although it only became well known in the nineteenth century (and, by the way, the notion that the apple fell on his head was a later embellishment introduced by Prime Minister Benjamin Disraeli's father, Isaac D'Israeli, in the early nineteenth century). The fullest account is by his friend William Stukeley (1687–1765), antiquarian and Stonehenge expert, who recalled a conversation in his own garden. Newton had, Stukeley wrote, been reminiscing about events nearly sixty years earlier, when as a student he had taken refuge from plague-riven Cambridge in the cottage of his birth at Woolsthorpe, a Lincolnshire hamlet. While he was sitting in the orchard, "the notion of gravitation . . . was occasion'd by the fall of an apple, as he sat in a contemplative mood. Why should that apple always descend perpendicularly to the ground, thought he to him self. Why should it not go sideways or upwards, but constantly to the earths centre? Assuredly, the reason is, that the earth draws it . . . there is a power, like that we here call gravity, which extends itself thro' the universe."[2]

Those of Newton's contemporaries who heard the story would have reacted very differently from how people do today. For one thing, the Bible was so important in their lives that they would immediately have thought about the Fall of Man in the Garden of Eden, when the serpent persuaded Eve to tempt Adam with a fruit from the forbidden Tree of the Knowledge of Good and Evil. This fruit became identified as an apple, probably because the Latin words for "evil" and "apple tree" are very similar—*malum* and *malus*. Medieval and Renaissance pictures of the infant Christ often show him holding an apple to symbolize that he is a second Adam, who will redeem fallen, sinful humanity. For his followers, Newton became a new Adam who would uncover God's mathematical laws of nature.

Furthermore, from a modern vantage point, the idea that the moon and an apple are subject to the same laws seems self-evident. But in the sixteenth and seventeenth centuries, many people still envisaged an Aristotelian universe, with its stark contrast between the eternal perfection of the heavens and the unruly chaos of the terrestrial sphere (see Myth 3). Opinions had already begun to shift, although it took around a century and a half for a satisfactory theory to emerge that corresponded to all the observations. Important discoveries were made by men like Nicolas Copernicus (1473–1543), who in 1543 suggested that the sun, not the earth, lies at the center of the universe, and Johannes Kepler (1571–1630), who by 1619 had formulated his three mathematical laws describing the elliptical orbits of the planets. Developing the research of these as well as many other predecessors, Newton united the cosmos into one single coherent entity by postulating a force of gravity that operated on earth as well as stretching out into the celestial regions.

Newton's sudden insight beneath the apple tree is the intellectual equivalent of a divine visitation, resembling the Apostle Paul's instantaneous conversion from skepticism to faith on the road to Damascus. According to their philosophical ideals, scientists proceed methodically, rather like Newton had himself—patiently accumulating evidence and ruthlessly testing hypotheses. The apple flatly contradicts that ideological vision, instead holding out the hope that years of painstaking research can be short-circuited by a blinding moment of insight. In the mid-nineteenth century, the new Oxford University Museum installed a statue of Newton dressed in a schoolboy's clothes and staring down at his apple, as if he were a born genius whose flash of inspiration had given him immediate insight into the truths of nature. Commissioned by the art critic and patron John Ruskin (1819–1900) to inspire students, this stereotype of innate brilliance implicitly minimizes the role of systematic scholarship (see also Myth 25).

This supposedly momentous event had no immediate impact. Whatever insights Newton may have gained from contemplating a falling apple, another twenty years went by before he published his theory of gravity.[3] During that time, he worked on several projects—including alchemy (see Myth 4) and optics—but returned to mathematical astronomy after several comets blazed unexpectedly across the sky. Although he repeatedly tested and modified his theories and experiments, physics might have continued unchanged if Newton's awestruck younger colleague Edmund Halley (1656–1742) had not chivied him into getting his masterpiece finished. Even when it was eventually published in 1687, the *Principia* did not immediately revolutionize knowledge. A reclusive scholar, Newton was uninterested in making his physics accessible, and his ideas were only gradually accepted, sometimes decades later.

It was not until the early nineteenth century that the story went public and Newton became celebrated as a scientific genius. During the eighteenth century, Newton's attribute was a small comet, symbolizing that by introducing precise prediction he had brought order to the universe and displaced astrologers as the experts of the heavens. The apple featured in a small French biography of 1821, which infuriated devotees by also insisting that Newton had experienced a bout of insanity, an unthinkable misfortune for a great British hero.[4] Despite this unpromising introduction, the apple rapidly became popular, and objections that this story undermined the Victorians' insistence on hard, patient work were soon ignored. The early nineteenth century was a time when history was being romanticized. Because of their allegorical significance, several myths were being established, such as the spider repeatedly weaving its web that supposedly inspired Robert Bruce (1274–1329) to rise up against the English invaders. Other freshly minted Newtonian tales included the disastrous occasion when his dog Diamond upset a burning candle on a second masterpiece, and a prearranged date during which the absent-minded professor tamped his pipe with the finger of a prospective wife. The apple was the one that survived to become a myth.

The No-God Fallacy

Modern scientists often point to Newton as the supreme Enlightenment rationalist—the world's first great physicist, who stamped out biblical superstition and replaced it with objective truth. In fact, he was a deeply religious man who dedicated his life to interpreting God's two great works: the Bible and the Book of Nature.

In Newton's version of Newtonianism, God is immanent throughout the universe and continuously intervenes in its welfare. In 1713, a quarter of a century after first publishing the *Principia*, Newton brought out a second edition. He added a few pages called the General Scholium, in which he outlined his views on the relationship between God and the created universe. God, wrote Newton there, "is Eternal and Infinite, Omnipotent and

Omniscient. . . . He endures for ever, and is every where present; and by existing always and every where, he constitutes Duration and Space."[5] Newton would have been appalled at the deterministic view now attributed to him, according to which the behavior of atoms is not affected by divine mandate but is governed by inexorable natural laws. This notion was introduced at the end of the eighteenth century by Pierre-Simon Laplace (1749–1827), the self-styled French Newton, who declared that if at any instant he knew where every atom was and how fast it was moving, he could—in principle, if not in practice—be sure of its whereabouts at any future time. This French Newton has acquired his own apocryphal story. "Where, pray, is God in your physics?" Napoleon Bonaparte (1769–1821) supposedly asked. "Sire," replied Laplace, "I have no need of that hypothesis." The original Newton had also shunned hypotheses, but for him God was no supposition but reality.

Newton has become an international figurehead of scientific reason, but he was a natural philosopher, a term that was emphatically not the early equivalent of a scientist. For natural philosophers, the whole point of studying nature was to discover more about God and his role as divine architect. Rather than using science to disprove holy texts, these students of nature validated experiments and theories by reconciling them with the scriptures. Turning to the Bible for guidance, Newton drew up plans of King Solomon's Temple, believing that its proportions reflected those of the universe itself. He bequeathed to the future a rainbow of seven colors not because he counted more accurately than his predecessors but because he believed that cosmic dimensions should follow the rules of musical harmony based on octaves. Following the mathematical precepts of Pythagoras (571–495 BCE), Newton searched for the numerical proportions binding God's creation together in harmonious perfection. As he wrote in the General Scholium, "And thus much concerning God; to discourse of whom from the appearances of things, does certainly belong to Natural Philosophy."[6]

Building on more than a century of innovation, Newton eventually brought together two contrasting approaches, indicated in the full title he chose for his great book, which translates from the Latin as *The Mathematical Principles of Natural Philosophy*. Whereas natural philosophers searched for fundamental explanations of how the world works, mathematicians focused on constructing descriptive models that might not necessarily represent reality but yielded useful results. Natural philosophers asked why things happen as they do; mathematicians wanted to know when, where, how much, and how often. By making gravity follow a simple mathematical relationship—the inverse square law—Newton emphasized that natural phenomena could be explained quantitatively, a fundamental shift in approach that proved crucial for modern science but did not eliminate God.

Corroborating the evidence from his published works, scientists as well as historians have written book after book making it clear how deeply engaged Newton was with theology, prophecy, numerology, alchemy, and other topics now viewed as having nothing whatsoever to do with science. Even so, other scientists persist in maintaining that Newton banished God from the cosmos. This willful blindness suits their own interests by enabling them to present science's past as ineluctable progress toward the truth. In this distorted vision of history, science emerges resplendent as a search for ultimate reality that relies on reason rather than ungrounded faith in unprovable entities. Reinforcing the status of scientific knowledge as incontrovertibly correct conveys the impression that scientific practitioners are inherently superior beings.

Fruit of Genius

Although different in kind, these two stories about Newton are linked by ideas about genius, one of those multilayered words that have shifted in meaning over the centuries. During Newton's youth, it was most commonly used to mean an innate gift or talent that had been imparted by God at birth. So, for instance, Newton was said to have a particular genius for mathematics, whereas women might have a genius for embroidery or singing. Gradually, the label moved toward describing an individual, but it was only in the early nineteenth century that the Romantic concept of a scientific genius emerged. Before then, "scientific genius" seemed a contradiction in terms: a genius's great poem or symphony arrives fully formed in his head (no possibility of a female genius), whereas a scientific thinker can explain every step along the way of his logical argument.

While Newton was alive, he was often twinned with the poet Alexander Pope (1688–1744) as great British emblems of Enlightenment. Partly under Newton's influence, experimental research began to rival literary composition as a worthwhile activity that would bring credit and profit to the nation; as Newton's status rose, so Pope's diminished. When Newton died in 1727, Pope composed this couplet, hoping (in vain) that it would be inscribed on Newton's tomb:

Nature and Nature's Laws lay hid in Night.
God said, Let Newton be! *and All was* Light.[7]

Echoing the biblical account of creation, Pope is here celebrating Newton as a scientific hero, a Christlike figure who has been sent by God to illuminate the darkness of ignorance and superstition on earth. The suddenness of the transition from confusion to knowledge is reiterated in the apple story, only this time inspiration arrives when Newton is already an adult—and whether the apple fell by chance or by divine design remains ambiguous.

As science and religion ostensibly separated during the nineteenth century, the role played by God in Newton's theories faded away, and genius took over the cultural significance and functions formerly attributed to sanctity. Tourists still flock to marvel at Woolsthorpe's decaying apple tree as if it were a saintly shrine, or stand in reverence before Newton's statue in the antechapel at Trinity College, Cambridge. His apple has become an iconic attribute resembling Daniel's lion or the wheel of St. Catherine of

Alexandria, and the Royal Society preserves locks of his hair as if they were saintly relics. Newton was undoubtedly an extremely clever man, but celebrating him as a scientific genius runs very close to worshiping him as a superhuman being.

Like the apple-tree myth, the no-God fallacy shows little sign of losing its grip, despite the solid arguments confirming its falsity. During the Victorian era, when technological science was becoming more powerful, Darwinism threatened many purpose-based models of evolution: according to Darwin's theory, the emergence of humanity was due to chance variation and natural selection, not the original plan of God the designer (see also Myth 11). Some scientists and clergymen found it advantageous to claim that science and religion were natural opponents, because that enabled both groups to carve out and maintain distinct yet high-status positions in society. At present, this rhetorical strategy is gaining new ground as a weapon in secular scientists' battle against creationism. Ironically, some of today's scientific militants are as dogmatic and demagogic as the religious fundamentalists they denounce so forcefully (see also Myths 13 and 24).

II

NINETEENTH CENTURY

THAT FRIEDRICH WÖHLER'S SYNTHESIS OF UREA IN 1828 DESTROYED VITALISM AND GAVE RISE TO ORGANIC CHEMISTRY

Peter J. Ramberg

The Wöhler synthesis is of great historical significance because for the first time an organic compound was produced from inorganic reactants. This finding went against the mainstream theory of that time called vitalism which stated that organic matter possessed a special force or vital force inherent to all things living. For this reason a sharp boundary existed between organic and inorganic compounds.

— *Wikipedia,* "Wöhler Synthesis" (2014)

In the early nineteenth century . . . many scientists subscribed to a belief that compounds obtained from living sources possessed a special "vital force" that inorganic compounds lacked. This notion, called vitalism, stipulated that it should be impossible to convert inorganic compounds into organic compounds without the introduction of an outside vital force. Vitalism was dealt a serious blow in 1828 when German chemist Friedrich Wöhler demonstrated the conversion of ammonium cyanate (a known inorganic salt) into urea, a known organic compound found in urine.

Over the decades that followed, other examples were found, and the concept of vitalism was gradually rejected. The downfall of vitalism shattered the original distinction between organic and inorganic compounds.

— David Klein, *Organic Chemistry* (2012)

In 1828, Friedrich Wöhler (1800–1882) published a short article in which he described the unexpected formation of urea from ammonium cyanate. The appearance of urea as a product was entirely unexpected, because theory predicted that cyanic acid and ammonia should produce a compound with the properties of a salt. Urea was not a salt, and it did not possess any of the properties expected for cyanates.[1] In the article, Wöhler repeatedly noted the novelty of the artificial synthesis, but he and his mentor, the well-known Swedish chemist Jöns Jakob Berzelius (1779–1848), were most intrigued by the formation of a nonsalt from a salt, and that ammonium cyanate and urea had the same elemental composition. Neither Wöhler nor Berzelius commented, as the epigraphs might suggest, on how the synthesis influenced the doctrine of vitalism, but within a few decades, chemists came to regard Wöhler's experiment as an "epochal" discovery that would mark both the death of vitalism and the birth of organic chemistry as a subdiscipline of chemistry. The myth has proved remarkably enduring—a survey of modern organic chemistry textbooks has revealed that 90 percent of them mention some version of the Wöhler myth.[2]

The urea myth can be conveniently condensed into three components: (1) that Wöhler synthesized urea from the elements, (2) that the synthesis unified organic and inorganic chemistry under the same laws, and (3) that the synthesis destroyed, or at least weakened, the idea of a "vital force" in living organisms. As historians have extensively documented, however, each of these three parts is highly problematic. First, Wöhler's synthesis could be, and was, rejected as artificial, because there may have been a residual "vital force" in his starting materials. Second, well before the urea synthesis, chemists had operated under the assumption, promoted by Berzelius, that organic and inorganic chemistry should follow the same laws of chemical combination. Third, "vitalism" was not a single theory but a variety of related ideas about the nature of life that continued well after Wöhler's synthesis in both chemical and biological contexts.

Inorganic Starting Materials?

For more than a century after Wöhler's synthesis, chemists and historians generally assumed that the synthesis was in fact "from the elements," and therefore completely artificial. In 1944, the chemist-historian Douglas McKie (1896–1967) argued that Wöhler's synthesis could never have sounded the "death-knell" for vitalism because his starting materials were actually derived from organic sources, and he had therefore not made urea directly from the elements. McKie concluded that "those who believe Wöhler drove vitalism out of organic chemistry will believe anything." According to McKie, the first total synthesis from the elements was achieved by Hermann Kolbe (1818–1885), who successfully made acetic acid from coal in 1845.[3] Until the mid-1960s, historians and chemists debated whether Wöhler had completed a "total" synthesis, but by 1970 it seemed clear that it was extremely difficult to establish whether or not Wöhler had made urea directly "from the elements."[4] The ambiguity surrounding the definition of "artificial" synthesis does not necessarily refute this component of the myth, but neither does it provide much support for it.

Unified Chemistry?

In the first decades of the nineteenth century, chemists were unclear about the nature of organic compounds and uncertain about whether they could be made artificially. This uncertainty does not mean, however, that chemists were without methods for investigating organic compounds as chemicals following the laws of chemical combination. Already in the 1780s Antoine Lavoisier (1743–1794) had demonstrated that compounds isolated from living things consisted mostly of the elements carbon, hydrogen, and oxygen. In the first decades of the nineteenth century, chemistry was strongly influenced by the newly discovered phenomenon of current electricity and the new theory of atomism developed by John Dalton (1766–1844). These two developments led

Berzelius to his theory of electrochemical dualism, in which inorganic compounds consisted of positively and negatively charged pieces held together by electrostatic attraction. Berzelius's theory worked well for inorganic compounds, and because they existed in a wide variety of simple combinations of many different elements, it was thought that the composition itself caused the chemical properties of the compound. But this theoretical framework did not seem to hold for organic compounds. Because carbon, hydrogen, and oxygen could combine to form so many different compounds, it appeared that composition alone could not account for the properties of each compound.

Despite these difficulties, chemists largely agreed that organic and inorganic compounds would follow definite laws of chemical combination. During the 1810s, for example, Michel Eugène Chevreul (1786–1889) closely studied the chemistry of naturally occurring fats and oils.[5] Chevreul separated animal fats into distinct compounds and determined that they had a definite composition, and that each fat consisted of a combination of glycerin with three large-molecular-weight fatty acids. Highly admired at the time, Chevreul showed that fats, like inorganic compounds, were subject to systematic chemical analysis and obeyed definite laws of chemical combination.[6]

Also in the 1810s, Berzelius became the champion of Daltonian atomism, consolidating the principles by which atomic combining proportions could be calculated. Berzelius had long thought that organic and inorganic chemistry should follow the same laws of combination, and in 1814 he turned explicitly toward organic chemistry, writing the following:

> It is evident that the existence of determinate proportions in inorganic bodies leads to the conclusion that they exist also in organic bodies; but as the composition of organic bodies differs essentially from that of those which are inorganic, it is clear that an essential modification must exist in the application of these laws to these two different classes of bodies.[7]

To show that organic compounds followed consistent laws of chemical combination, Berzelius developed techniques to determine the proportion of carbon, oxygen, and hydrogen atoms in pure organic compounds, finding that each consistently gave the same proportion of elements. He concluded that organic compounds could be interpreted in terms of the atomic theory and had definite measurable combining proportions, as did inorganic compounds.[8]

In his comprehensive and influential book *Essay on the Theory of Chemical Proportions* (1819), Berzelius argued simply that the *combining proportions* of the elements in organic compounds were more complex than were those in inorganic compounds.[9] Believing that inorganic chemistry could serve as an analogy for understanding the composition of organic compounds, Berzelius argued that his law of electrochemical dualism should also apply to organic compounds.[10] The principal stumbling block for the synthesis of organic compounds was therefore not ignorance of a different kind of chemical force that held organic compounds together but the complexity of the "arrangement" of atoms in the molecule.

The Demise of Vitalism?

In most versions of the myth, vitalism is assumed to be a theory that supposes the existence of a mystical, nonmaterial entity that is present in living things but absent in inorganic systems—a "rational soul" that is responsible for maintaining the complex systems found in living organisms. This is certainly one version of vitalism, defended, for example, by Georg Ernst Stahl (1659–1734) early in the eighteenth century. But Stahl's version of vitalism is one extreme on a continuum of ideas about the nature of life. At the opposite extreme lay a pure materialism, in which living things are envisioned as complicated machines governed solely by physical and chemical laws. This completely "anti-vitalistic" position was already well established by the mid-eighteenth century, held most famously by Julien Offray de La Mettrie (1709–1751) in his book *Man a Machine* (1748).[11]

Other natural philosophers staked out different conceptions of vitalism. Albrecht von Haller (1708–1777) and Xavier Bichat (1771–1802), for example, rejected the concept of vitalism as an external nonmaterial entity, suggesting instead that living things possessed certain kinds of forces, analogous to Newtonian gravitational attraction, which could be characterized and studied but whose ultimate nature remained unknown. Johann Friedrich Blumenbach (1752–1840) and Johann Christian Reil (1759–1813) developed yet another version of vitalism, called "vital materialism," in which the vital force was not an independent entity but something that emerged from the complex interaction of the chemical and physical components of the organism.[12] Although living systems were governed by chemical and physical laws, they were more than simply the sum of their parts. These examples show that vitalism was not a single, comprehensive theory but a variety of theories about biological systems.

Berzelius himself had developed a version of vital materialism as early as 1806, which he incorporated into the section on organic chemistry in the 1827 edition of his textbook; he never substantially revised the entry in subsequent editions.[13] Similarly, Justus von Liebig (1803–1873) described vital force in *Animal Chemistry* (1842) as "a peculiar property, which is possessed by certain material bodies, and becomes sensible when their elementary particles are combined in a certain arrangement or form."[14] This force, analogous to gravity or electricity, arose from the complexity of the system. Because the synthesis of a single compound could have had little effect on vitalistic theories about an organized system, it should not be surprising that Wöhler and Berzelius failed to discuss the impact of the synthesis of urea on vitalism in their correspondence and that early textbooks on organic chemistry did not mention Wöhler or the urea synthesis.[15]

Among biologists, vitalism continued to wax and wane. In the 1890s, Hans Driesch (1867–1941) suggested the existence of an explicitly nonmaterial factor he named "entelechy," which directed the growth of the organism; and during the 1920s, Nobel Laureate Hans Spemann (1869–1941) developed a sophisticated holistic theory of embryonic development that was commonly mistaken by his contemporaries as vitalistic.[16] The modern concepts of "emergent structures" and "self-organization" to explain such diverse phenomena as biological systems, the origin of life, intelligence, as well as economic systems and galaxy formation, are themselves reincarnations of vital materialism.[17]

Even the increasing success of chemists in making organic compounds did not completely eliminate the gap between organic and inorganic compounds. In 1848, Louis Pasteur (1822–1895) suggested that molecules of tartaric acid could exist in left- or right-handed asymmetric forms, leading to the conclusion that many compounds isolated from living things consisted of only one of these two forms. Attempts to create them artificially resulted in both forms without preference. In his famous 1860 lecture "On the Asymmetry of Natural Organic Products," Pasteur declared that the presence of asymmetry was "perhaps the only well marked line of demarcation that we can at present draw between the chemistry of dead nature and the chemistry of living nature."[18] According to Pasteur, asymmetric molecules could be produced only by "asymmetric forces," a view closely related both to his conviction that alcoholic fermentation was a vital and not a chemical process, and to his later attempts to disprove spontaneous generation (see Myth 15).[19]

In 1898, the British chemist Frances Japp (1848–1928) argued explicitly that asymmetry at the molecular level required a nonmaterial cause:

> At the moment when life first arose, a directive force came into play—a force of precisely the same character as that which enables

the intelligent operator, by the exercise of his Will, to select one crystallized enantiomorph and reject its asymmetric opposite.[20]

In 1894, Emil Fischer (1852–1919) suggested an unequivocally chemical and mechanistic view of fermentation and enzyme action, in which the source of biological asymmetry was the asymmetry already present in the enzymes that fit asymmetric molecules like a lock and key. Japp deftly countered Fischer by noting that even if this mechanistic interpretation were true, it still left unexplained the origin of asymmetry. Indeed, the origin of molecular asymmetry remains an unsolved problem today.

Reasons for Endurance

Like other myths in this volume, the Wöhler myth shows no signs of fading away, because it serves several specific purposes. For organic chemists, it provides a hero who accomplished a specific datable task that assumed great significance. The myth became widespread after Wöhler's death in 1882, in part to validate the theoretical autonomy of organic chemistry as a discipline that no longer required concepts from either biology or physics, and in part because German chemists wished to place the origins of the powerful German chemical community, in which synthesis played a central role, squarely in their own country.[21] For biologists, the myth's simplistic image of vitalism provides a convenient foil for depicting how physiologists adopted the rigorous mechanistic and quantitative methods of chemistry and physics in the process of making biology more "scientific" by ridding it of "pseudoscientific" entities such as vital forces.[22]

THAT WILLIAM PALEY RAISED SCIENTIFIC

QUESTIONS ABOUT BIOLOGICAL ORIGINS

THAT WERE EVENTUALLY ANSWERED BY

CHARLES DARWIN

Adam R. Shapiro

[William Paley's] . . . *Natural Theology—or Evidences of the Existence and Attributes of the Deity Collected from the Appearances of Nature*, published in 1802, is the best known exposition of the "Argument from Design," always the most influential of the arguments for the existence of a God. It is a book I greatly admire, for in its own time the author succeeded in doing what I am struggling to do now. . . . The only thing he got wrong—admittedly quite a big thing!—was the explanation itself. He gave the traditional religious answer to the riddle, but he articulated it more clearly and convincingly than anybody had before. The true explanation is utterly different, and it had to wait for one of the most revolutionary thinkers of all time, Charles Darwin.

—Richard Dawkins, *The Blind Watchmaker* (1986)

Up until the time of Darwin, in fact, the argument that the world was designed was commonplace in both philosophy and science. But the intellectual soundness of the argument was poor, probably due to lack of competition from other ideas. The pre-Darwinian strength of the design argument reached its zenith in the writings of nineteenth-century Anglican clergyman William Paley. An enthusiastic servant of his God, Paley brought a wide scientific scholarship to bear on his writings but, ironically, set himself up for refutation by overreaching.

—Michael Behe, *Darwin's Black Box* (1996)

Richard Dawkins and "intelligent design" advocate Michael Behe
don't agree on much, certainly not when it comes to evolution.
They can't even agree on whether William Paley (1743–1805)
should be seen as a man of the eighteenth century or the nineteenth. But the one point of consensus that they've reached is
historical, claiming that Charles Darwin's (1809–1882) *On the
Origin of Species* (1859) refuted Paley's *Natural Theology* (1802).
They agree that Paley was trying to explain the origin of complex
structures that we see in the natural world—things such as eyes,
ears, lungs, wings, and other organs—which seem to serve purposes like that of a watch telling time. It was this need to explain
biological origins that led Paley to conclude that a deity—an intelligent designer—must exist; and it was not until Darwin published his theory of evolution by natural selection that Paley's
argument was seriously challenged.

This myth—that Darwin refuted Paley's scientific explanation
for the origins of complex life—is mistaken on several levels. Paley's argument wasn't about biological *origins,* and it wasn't a
scientific argument but a theological one. While Darwin eventually found Paley's arguments to be unconvincing, his goal wasn't
to refute them. In fact, some of Darwin's contemporaries found
natural selection to be compatible with natural-theology arguments similar to Paley's. It was not until the twentieth century,
when a new synthesis of evolutionary thought was held against
a caricature of Paley's original ideas, that people began to claim
that a scientific argument for design had been refuted.

The *Natural Theology* begins with Paley stating that if he were
to come across a stone in a field, "I might possibly answer, that,
for anything I knew to the contrary it had lain there for ever." By
"for ever," he didn't mean since the beginning; he meant *eternally,*
without beginning or end. At the time that Paley wrote the *Natural Theology,* the idea of an eternal universe, one that had no
beginning or end in time, was given serious consideration by astronomers and geologists; it had also been a source of Christian
theological debate for centuries. An infinite eternal universe is not

ruled out when Paley then suggests how we might react were we to come across a watch instead of a stone. It doesn't matter how the watch first came to be there; what matters is that the watch shows evidence of having a purpose at the moment he's looking at it. Even though it operates by springs and not by gravity (the way hourglasses, pendulum clocks, or descending-weight clocks do), its parts are arranged in such a way that it coincides with the way that gravity makes the planets move—in other words, it measures the time. It's as if, Paley writes, one being was responsible for the laws of nature, and another for putting together material objects in the world. Lo and behold, those material objects (eyes, ears, lungs) seem ready to take advantage of—are *adapted to*—the laws of nature (optics, acoustics, pressure). It's the fact that material objects are adapted to this world, with its specific natural laws, that leads Paley to conclude that there's a designing intelligence behind it all.[1]

This argument doesn't depend on how those material objects were formed in the first place. Paley invites the reader to imagine a watch that, as part of its mechanical function, assembles another watch just like itself. We might then expect that the watch in front of us now was formed by a previous watch, and that one by another (again imagining a world that has existed forever). His argument doesn't change; he examines the watch for its evidence of purpose, not to ask how it was initially formed.[2]

Paley did believe that the world had a beginning, but he didn't want to take that for granted. Thus, he restricted his argument for a designer, so that it would be true even for an eternal world. This mattered because his intention was never simply to show that some designer merely existed. His aim was to do theology: to answer religious questions about the kind of being that could create a world filled with purpose. Because natural laws apply everywhere and act in the same ways, he inferred that there is only one designer and that that designer is everywhere: it is God. Because (as he saw it) the universe contains no unnecessary suffering, and the experience of pleasure often seems to be an end in itself, he inferred that the designer is good. And because the world gives evidence of God that we can explore through studying nature and using our rational minds, he inferred that God wants all people to be able to understand him, even before they accept revelation, scripture, or some other private way of knowing.[3]

One could try to prove that a god exists by saying that there are things that nature can't explain, that there are phenomena that couldn't have occurred through natural processes alone. But Paley feared that such a proof could lead to appeals to private knowledge and would justify religious conflict between different interpretations of revelation or scripture. Paley invoked nature as the starting point for theology because he thought that public knowledge based on observation of nature was the best hope for consensus.[4]

For Paley, the natural world testified that the best society was a conservative one, in which people followed their natural God-given inclinations, and that society as a whole had an optimal balance of intellectuals, laborers, governors, craftsmen, and soldiers. When people disregarded those natural inclinations (Paley thought that these were mostly hereditary), this divine utilitarian system—one that brought the most benefit to the society as a whole—broke down. In Paley's view, that was precisely what happened in the French Revolution.[5]

For Paley, nature is God's way of illustrating morality. Not just the parts of organs but whole ecosystems are balanced to bring about the least amount of suffering. Paley used nature—the structures and organs of people and animals and the relationship among different species—to make claims about God and God's moral law. Paley wasn't engaged in science, and not just because it would be somewhat anachronistic to use the word "science" for these descriptions of nature. Paley used natural history and natural philosophy instead, not in order to reach conclusions that we would today call scientific but in order to make religious claims. Paley didn't invoke God to explain nature; he invoked nature to explain God. The *Natural Theology* was, truly, a work of theology.[6]

By the 1830s, the presentation of Paley's argument had already changed. In an annotated edition of the text published in 1836, editors Henry Brougham and Charles Bell added a footnote to the first paragraph, stating that with recent discoveries in geology, it was no longer true that we should infer that the rock had lain there forever—it, too, had been formed at some distinct time. While this footnote added some updated geological knowledge, it undermined Paley's theological point: that a designer was not dependent on origins. With new theories about the development of the solar system and geological formation, an eternal universe was of less concern to readers in the 1830s. For them, the real question was not whether the world had a beginning but whether (or how) things had changed since that beginning.[7]

This is where the question of the origin of species comes in. Charles Darwin was not the first to suggest that life on earth had changed over time. When Paley wrote in 1802, he considered two primary theories for the introduction of new species. First was the theory of Georges-Louis Leclerc, Comte de Buffon (1707–1788), whereby the world is suffused with creative particles called *organic molecules* that have an inherent tendency to organize matter and give rise to new life. Buffon's theory wasn't truly evolutionary because it didn't call for gradual emergence, but it did claim that new species came about at some time after the initial creation of the world. The other theory Paley addressed was that of *appetencies*, put forward by Erasmus Darwin (1731–1802), Charles's grandfather, in which life was seen as the process of willful striving for new growth and new complexity. This was a truly evolutionary theory, which Paley described as new forms emerging "by continual endeavors, carried on through a long series of generations."[8]

Paley firmly rejected Buffon. Calling his theory both incorrect and atheistic, Paley stated that there was no evidence that new species had ever spontaneously emerged, nor was there any backing for the hypothesis that nature has the inherent power to create new species this way. But he treated Erasmus Darwin's theory of appetencies differently. According to Paley, there was very little evidence (in 1802) to support the claim that evolution happened in the past, but it was not impossible that such change *could have* occurred. "I am unwilling to give to it the name of an *atheistic* scheme," he wrote, because "the original propensities and the numberless varieties of them . . . are, in the plan itself, attributed to the ordination and appointment of an intelligent and designing Creator." Unlike Buffon, Erasmus Darwin didn't posit any additional creative force in the world. The natural life processes of growth and inheritance, gradually accumulating over the course of many generations, could give rise to new variation. But Paley still objected that Erasmus Darwin's particular version of gradual evolution "does away with final causes." Paley wasn't an evolutionist, but his argument didn't rely on species being specially created. Because his argument for a designer came from indications of purpose, not origin, Paley did not insist that an evolutionary scheme was incompatible with his argument (as long as there was still a role for purposes).[9]

Charles Darwin posited a system whereby species evolve over time, not because they have some innate creative power but because those more adapted to the demands of their environment tend to be better at surviving and reproducing. Natural selection does not eliminate purposes; it serves a purpose of increased (but never complete) adaptation. Darwin argued that natural selection brought about an overall good for creatures in a way similar to Paley's utilitarianism. Indeed, the only time Paley is mentioned in the *Origin of Species,* it's not in refutation but in approval:[10]

> Natural selection will never produce in a being any structure more injurious than beneficial to that being, for natural selection acts solely by and for the good of each. No organ will be formed, as Paley has remarked, for the purpose of causing pain or for doing an injury to its

possessor. If a fair balance be struck between the good and evil caused by each part, each will be found on the whole advantageous.[11]

Even though Darwin said that he eventually found Paley's religious conclusions unconvincing, he never saw them as scientific arguments about the origin of life that had to be refuted in order to make the case for evolution. It wasn't until the twentieth century, after the advent of a neo-Darwinian synthesis that explicitly tried to eliminate any talk of purpose or progress from biology, that Paley was resurrected as a caricature of pre-Darwinian biological thought. Paley was virtually ignored at the time of the 1925 Scopes antievolution trial, when religious opposition to evolution was presented as a conflict with the Bible, not with natural theology. Historians and other commentators at the time of the 1959 centenary of the *Origin* mentioned Paley in contrast to Darwin and in ways that earlier texts had not. By the 1980s, accounts of the Paley-Darwin relationship (like that given by Dawkins) made the *Natural Theology* out to be a work of science rather than religion.[12]

THAT NINETEENTH-CENTURY GEOLOGISTS WERE DIVIDED INTO OPPOSING CAMPS OF CATASTROPHISTS AND UNIFORMITARIANS

Julie Newell

We may assert, with our author [Charles Lyell] and other geologists, that all the facts of geological observation are *of the same kind* as those which occur in the common history of the world. The question then comes before us,—are the extent and the circumstances of the geological phenomena *of the same order* as those of which the evidence has thus been collected? Have the changes which lead us from one geological state to another been, on a long average, uniform in their intensity, or have they consisted of epochs of paroxysmal and catastrophic action, interposed between periods of comparative tranquillity?

These two opinions will probably for some time divide the geological world into two sects, which may perhaps be designated as the *Uniformitarians* and the *Catastrophists*.

—William Whewell, review of *Principles of Geology* (1832)

Men are born either catastrophists or uniformitarians. You may divide the race into imaginative people who believe in all sorts of impending crises,—physical, social, political,—and others who anchor their very souls *in statu quo*. There are men who build arks straight through their natural lives, ready for the first sprinkle, and there are others who do not watch Old Probabilities or even own an umbrella. This fundamental differentiation expresses itself in geology by means of the two historic sects of catastrophists and uniformitarians.

—Clarence King, "Catastrophism and Evolution" (1877)

Geology is a *historical* science, and as such, it focuses much of its work on creating an accurate and detailed account of the past. Understanding the past depends on finding evidence, figuring out what actually counts as evidence, and understanding how that evidence fits together to tell the story of past events. For the science of geology to develop, geologists had to agree on valid methods for doing all of these things. This involved a great deal of observation in many different places and a series of lengthy arguments about how to interpret those observations. These activities produced lively and detailed arguments among educated and experienced men. Accounts of these arguments tend to portray participants as members of opposing camps, drastically oversimplifying and otherwise misrepresenting their ideas.

The disagreement in the late eighteenth century between the Wernerians and the Huttonians provides an early example. Wernerians supposedly agreed with Abraham Gottlob Werner (1749–1817) that the earth's rock layers formed by precipitation from the ocean. Similarly, Huttonians supposedly accepted the argument of James Hutton (1726–1797) that subterranean heat played the primary role in forming the earth's rock layers. Both Hutton and Werner presented detailed ideas involving a great deal more than fire versus water. Many of those labeled "Huttonians" or "Wernerians" were much more interested in the practical utility of the respective theories than in adhering strictly to a theoretical system of rock formation.

Hutton's ideas did not become known through his own writings, published in very limited numbers and full of long passages of description that made them very difficult to read.[1] John Playfair's (1748–1819) *Illustrations of the Huttonian Theory of the Earth* (1802) made Hutton's ideas much more available to readers. Playfair referred to yet another contrasting set of names:

> It is foreign from the present purpose, to enter on any history of the systems that, since the rise of this branch of science, have been invented to explain the phenomena of the mineral kingdom. It is suf-

ficient to remark that these systems are usually reduced to two classes, according as they refer to the origin of terrestrial bodies to FIRE or WATER; and that, conformably to this division, their followers have of late been distinguished by the fanciful names of *Vulcanists* and *Neptunists*. To the former of these DR HUTTON belongs much more than to the latter; though, as he employs the agency of both fire and water in his system, he cannot, in strict propriety, be arranged with either.[2]

In reality, neither could anyone else. Humans rarely sort perfectly into either-or categories, and especially not when it comes to their ideas.

Three of Hutton's central concepts subsequently influenced geological thinking in profound ways:

Actualism: geological change must be explained by mechanisms observable in the present—that is, *actual* causes.

Gradualism: geological explanations must be limited not only to the kinds of forces now observed (actualism) but also to the observed rates at which they act.

Time: actualism and gradualism together require vast amounts of time to explain the observable geological record.

All of these ideas played a central role in Charles Lyell's (1797–1875) three-volume *Principles of Geology,* one of the most famous publications in the history of geology. The concepts of actualism and gradualism appear in the very title of the first volume, published in 1831: *Principles of Geology, Being an Attempt to Explain the Former Changes of the Earth's Surface, by Reference to Causes Now in Operation.* Lyell insisted that geological change in the past must be explainable by relying on the same *kinds* of geological processes observable in the world today, such as erosion and deposition by water and eruption of volcanoes. He also insisted that these forces always acted at the same *degree* of intensity observed today, which would have required immense periods of time to produce the observed effects. Lyell used the rhetorical strategy of uniting actualism and gradualism into the single concept of uniformity to make the concepts seem inseparable.[3]

Fellow geologist and clergyman William Daniel Conybeare (1787–1857), reviewing Lyell's first volume, commented that "no real philosopher, I conceive, ever doubted that the physical causes which have produced the geological phænomena were the same in kind, however they may have been modified as to the degree and intensity of their action, by the varying conditions under which they may have operated at different periods." Essentially, Conybeare argued that few geologists disagreed with Lyell's insistence on known geological causes, but most believed existing causes would operate differently under different conditions. Claims about what happened under unknown past conditions were unqualified speculation, not sound science. Uniformity combined the sound principle of actualism with the speculative principle of gradualism.[4]

In his 1832 review of the second volume of Lyell's *Principles* (1832), philosopher and historian of science William Whewell (1794–1866) introduced yet another either-or categorization of geological thinkers: uniformitarians and catastrophists. Although these categories have been used ever since, they are seldom clearly defined or carefully applied. Whewell, like Conybeare, insisted that the *rate* of geological change in the past was the key point of disagreement between Lyell—the uniformitarian—and most other geologists.[5]

As demonstrated in the epigraph, Whewell chose the term "catastrophist" for the majority of geologists, who insisted that the causes of geological change might have proceeded at different rates during different periods of the earth's history. Lyell argued that to accept phenomena of a different order was speculative and thus poor science. Others (Whewell and Conybeare among them) argued that Lyell's rejection of the possibility that the rate at which forces acted might vary over time was speculative and thus

poor science. On many other important points, the so-called catastrophists and uniformitarians agreed and even praised each
other's work.[6]

The equation of uniformitarianism with Lyell and "science,"
and catastrophism with supposedly faith-based anti-Lyellian
positions both oversimplified and misrepresented the ideas
and arguments of Lyell's scientific critics. Lyell himself laid the
groundwork for this view of his opponents in his history of
geology, which made up a large portion of the first volume of
Principles. Whewell's choice of the term "catastrophist" facili-
tated conflation of geologists who rejected Lyell's insistence on
gradualism with other writers who invoked scriptural authority
and supernatural causes to explain geological observations. The
former were well-established scientists, many of them, such as Conybeare, men of both faith and science. The latter wrote primarily
for a broad, popular audience and were men of faith but not science. The conflict existed not between science and religion but between those who rejected and those who accepted extrascientific
sources of information and causes in geological explanation.[7]

Lyell's *Principles* was widely known and read. It was engagingly
written and illustrated, frequently updated in new editions, and
affordable. Lyell had great incentive to keep *Principles* accessible and up to date because, having abandoned the practice of
law, he relied largely on the income from sales of his books.
Principles went through twelve British editions and at least one
American edition (1842). The influence of some of Lyell's ideas
on Charles Darwin (1809–1882) extended the fame and impact
of *Principles*.[8]

Like their British colleagues, American geologists read, debated, and selectively adopted Lyell's powerful ideas. In the United
States, men of faith were prevalent among the leading geologists
of the period, including Benjamin Silliman (1779–1864), Edward Hitchcock (1793–1864), and James Dwight Dana (1813–
1895). Their faith increased the appeal of their science for much
of their American audience but did not interfere with the sound-

ness of their scientific work. Their writings represent the very antithesis of a conflict between science and religion.[9]

By the beginning of the twentieth century, *uniformitarian* had come to refer to a mythical practitioner of true science, opposed to the supposed interference of religion in interpreting the history of the earth and symbolized by Charles Lyell and his *Principles of Geology*. *Catastrophist* had come to symbolize geologists whose understandings were influenced by religious sources and supernatural forces. By the beginning of the twenty-first century, "catastrophic" cause—such as the impact of extraterrestrial bodies on the earth and failed ice dams that released previously unimaginable floods—had became well-substantiated natural forces in the scientific explanation of the geological history of the planet. A growing literature portrays modern catastrophists as careful scientists guided by the evidence while implying that uniformitarians have been blinded by their scientific dogma.[10]

Since Whewell coined "uniformitarian" and "catastrophist" in 1832, both terms have been more about identifying good guys and bad guys than about understanding the issues involved or the ideas of individual geologists. Neither term has been clearly defined nor consistently applied. They are neither camps into which geologists usefully divided themselves nor analytic categories contributing to understanding the history of geology.

THAT LAMARCKIAN EVOLUTION RELIED LARGELY ON USE AND DISUSE AND THAT DARWIN REJECTED LAMARCKIAN MECHANISMS

Richard W. Burkhardt Jr.

A rival theory [to Darwin's theory of natural selection], championed by the prominent biologist Jean-Baptiste Lamarck, was that evolution occurred by the inheritance of acquired characters. According to Lamarck, individuals passed on to offspring body and behavior changes acquired during their lives. . . . In Darwin's theory, by contrast, the variation is not created by experience, but is the result of preexisting genetic differences among individuals.

—Peter H. Raven and George B. Johnson, *Biology* (2002)

When biology textbooks compare the evolutionary theories of Jean-Baptiste Lamarck (1744–1829) and Charles Darwin (1809–1882), they commonly lead readers to believe (1) that the inheritance of acquired characters was the primary mechanism of Lamarck's theory, and (2) that Darwin rejected this mechanism and put natural selection in its place. To cite but one example, the preceding quotation comes from one of the best texts in the business.

There are two fundamental problems with accounts such as this. First, the inheritance of acquired characters—that is, the inherited effects of use and disuse—did *not* constitute the primary factor of Lamarck's theory of organic change. Second, Darwin believed firmly in the inheritance of the effects of use and disuse, allowing that it aided natural selection "in an important manner."[1] What is more, whereas Lamarck never offered an explanation of how characters acquired through the effects of use or disuse in one generation could be passed on to the next generation, Darwin did so in his "provisional hypothesis of pangenesis."[2] To top things off, Darwin toward the end of his career confidently suggested that he had assembled more observations on the inherited effects of use and disuse than had any other author.[3]

Let us begin with the case of Lamarck. It is true that use inheritance was an important part of his evolutionary thinking, but Lamarck in fact offered a *two*-factor theory of evolution, a theory in which use inheritance appeared as the *secondary* factor, not the premier, predominant one. Use inheritance was key to Lamarck's discussions of change at the species level, but his theory of organic change was broader than that. It represented Lamarck's effort to construct a general picture of the mutual affinities of the diverse forms of animal life and then explain why that picture looked the way it did.[4]

As professor of invertebrate zoology at the Museum of Natural History in Paris, Lamarck sought to bring order to the dizzying array of forms that Carl Linnaeus (1707–1778) had unsatisfactorily crowded together into two animal classes: insects and worms. Following the lead of Antoine-Laurent de Jussieu (1748–1836) in botany and Georges Cuvier (1769–1832) in zoology, Lamarck decided that the best way to identify which organisms were most similar to each other (and, alternatively, most unlike each other) was to consider their internal organs rather than their external forms. He found that distinguishing the organ systems of the different animals not only allowed him to differentiate the broad classes of animals from one another but also revealed that these classes could be ordered in a single series of increasing or decreasing complexity, depending on whether one started with the least complex or the most complex types. As of 1800, he had distinguished seven classes of invertebrates— polyps, radiates, worms, insects, arachnids, crustaceans, and mollusks—to go along with the four classes of vertebrates—fish, reptiles, birds, and mammals. By 1815 he would introduce half a dozen more invertebrate classes.[5]

The year 1800 was also when Lamarck began telling his students about his new thoughts on organic mutability. Most of what he initially told them related to change at the species level, but he also hinted at a much broader view wherein nature began with the very simplest forms of life and, from these, successively produced all the rest. Two years later he made the latter point more forcefully, urging his students to

> ascend from the simplest to the most complex; leave from the most imperfect animalcule and go up the length of the scale to the animal richest in organization and faculties; conserve everywhere the order of relation in the masses; then you will have hold of the true thread that ties together all of nature's productions, you will have a just idea of her order of proceeding, and you will be convinced that the simplest of her living productions have successively given rise to all the others.[6]

Over the next decade and a half, as he elaborated on how the diverse forms of life came into being, Lamarck insisted that the tendency toward increased complexity in the animal scale took priority over competing factors—factors that prevented the scale from being completely regular. In his *Zoological Philosophy* of 1809 he wrote, "The state in which we now see all the animals is on the one hand the product of the increasing composition of organization, which tends to form a regular gradation, and on the other hand that of the influences of the multitude of very different circumstances that continually tend to destroy the regularity in the gradation of the increasing composition of organization."[7] Six years later, in the first volume of his great treatise, *The Natural History of the Invertebrates,* he referred to "two very different causes" of animal diversity. Of these, "the first and predominant cause" gave animal life the power of progressively complicating animal organization. The second cause, unrelated to the first, was the influence of diverse environmental circumstances. This cause was responsible for the gaps in the series, for the branches that diverged from the series and made it less simple, and for certain anomalies in the different organ systems.[8]

Lamarck attributed the tendency to increased complexity to what he called "the power of life," which he explained in wholly materialistic terms as the hydrodynamic action of fluids moving through the animal body, carving out channels and organs, and gradually, over immense periods of time, making animal organization increasingly complex.[9] The influence of environmental circumstances on the living body, by Lamarck's account, was more indirect. Animals develop new habits, he explained, in responding to differences of location, temperature, climate, and so on. In adopting new habits, they end up using some organs more and some organs less. This, in turn, modifies the animals' faculties and structures in ways that allow them to "conserve and propagate themselves by generation."[10]

Lamarck presumed that both modifications produced in the animal body by the creative action of fluids and modifications acquired as the result of the use or disuse of parts were passed down to successive generations. It was in connection with the inherited effects of use and disuse, however, that he offered the particular examples that have since become so closely associated with his name, such as that of the giraffe acquiring its long neck and long forelegs as the result of generations of members of the species stretching upward to browse on tree leaves. This, too, was the focus of the "two laws" that he announced in his *Zoological Philosophy* and that have since been too often understood to represent the whole of his theorizing:

First Law: In every animal that has not gone beyond the period of its development, the more frequent and sustained use of any organ strengthens this organ little by little, develops it, enlarges it, and gives to it a power proportionate to the duration of this use; while the constant disuse of such an organ insensibly weakens it, deteriorates it, progressively diminishes its faculties, and finally causes it to disappear.

Second Law: All that nature has caused individuals to acquire or lose by the influence of the circumstances to which their race has been exposed for a long time, and, consequently, by the influence of a predominant use of such an organ or constant disuse of such a part, she conserves through generation in the new individuals that descend from them, provided that these acquired changes are common to the two sexes or to those which have produced these new individuals.[11]

Significantly enough, Lamarck did not claim the idea of the inheritance of acquired characters as his own. Nor did he feel any need to gather evidence or conduct any experiments in support of this idea or to hazard a hypothesis regarding how acquired characters might be transmitted from one generation to the next. In 1815 he allowed that "this law of nature" was "so true, so striking, so much attested by the facts, that there is no observer who has been unable to convince himself of its reality."[12]

To acknowledge that the idea of use inheritance was common enough in Lamarck's day is not to say, however, that Lamarck was in no way original in embracing it. The other writers who believed in use inheritance typically denied that it could proceed so far as to produce species change.[13] Where Lamarck claimed originality and priority was in recognizing "the importance of this law and the light that it sheds on the causes that have led to the astonishing diversity of animals." He took greater satisfaction in this accomplishment, he said, than in all of the distinctions he had made in forming "classes, orders, many genera, and a quantity of species"—the kind of work, he said, that other zoologists took as the sole object of their studies.[14]

Let us turn now to Darwin. As indicated previously, Darwin was a strong believer in the inheritance of acquired characters, notwithstanding his conviction that natural selection was the primary agent of organic change. In the first edition of his *On the*

Origin of Species (1859), he stated, "I think that there can be little doubt that use in our domestic animals strengthens and enlarges certain parts, and disuse diminishes them; and that such modifications are inherited."[15] He noted that the domestic duck, for example, compared to the wild duck had smaller wings but bigger leg bones, apparently the result of flying less and walking more than its wild counterpart. Likewise, the drooping ears of many domestic animals, as contrasted with the upright ears of the animals' wild relations, seemed attributable to "the disuse of the muscles of the ear, from the animals not being much alarmed by danger." Similarly, he attributed the blindness of certain cave animals "wholly to disuse," while he suggested that the rudimentary eyes of moles and certain other burrowing rodents represented the effects of disuse "aided perhaps by natural selection."[16]

Interestingly enough, Darwin felt no need to attribute the idea of use inheritance to Lamarck. Lamarck had called on the acquired effects of disuse to explain why domesticated ducks could not fly as well as wild ducks could and why moles had only rudimentary eyes, but Darwin did not cite him in this regard.[17]

Throughout his career, Darwin sought to distance his ideas from Lamarck's. He felt that Lamarck had not made a good case for evolution, that Lamarck's notion of a tendency to increased complexity lacked foundation, and that Lamarck had been wrong to suggest that animals acquired new organs by wanting or wishing for them (a charge to which Lamarck had opened himself through careless writing, though it did not in fact correspond to his general views). It may be that the only time Darwin explicitly connected the idea of use inheritance with Lamarck's name was in an instance in which Darwin was confident that use inheritance could not be effective, but natural selection could. Writing about the specialized instincts and structures of neuter castes of insects, Darwin observed, "No amount of exercise, or habit, or volition, in the utterly sterile members of a community could possibly have affected the structure or instincts of the fertile members, which alone leave descendants." He then went on to say, "I am surprised that no one has advanced this demonstrative case of neuter insects, against the well-known doctrine of Lamarck."[18]

That said, Darwin believed that use inheritance was a very real phenomenon. One of the goals of his 1868 work, *The Variation of Animals and Plants under Domestication,* was to explain "how the effects of the long-continued use or disuse of any part, or of changed habits of body or mind, can be inherited." Remarking that "a more perplexing problem can hardly be proposed," he was pleased to think he had found a solution. In his "provisional hypothesis of pangenesis," he postulated that the cells of the body threw off hereditary particles ("gemmules"), which were transmitted via reproduction to the next generation. He proposed that when body cells were modified by use or disuse, they produced gemmules that were modified accordingly.[19]

In 1872, in the sixth edition of the *Origin,* Darwin represented natural selection as the chief agent of organic change, but he singled out "the inherited effects of the use and disuse of parts" as a factor that aided natural selection "in an important manner"— as opposed to various other factors that acted (with respect to adaptation at least) in "an unimportant manner." In doing so, he complained about those who had misrepresented his views by stating that he had "attribute[d] the modification of species exclusively to natural selection," thereby ignoring all he had said about the importance of the use and disuse of parts. With a testiness that was unusual for his published writings, he observed, "Great is the power of steady misrepresentation; but the history of science shows that fortunately this power does not long endure."[20]

The present chapter attests—as does the whole of this volume— that the first part of Darwin's statement about the power of steady misrepresentation was right but that the second part was overly optimistic. Despite the efforts of historians of science to lay them to rest, many common myths about the history of science have endured. Included among these are the myths that the inherited effects of use and disuse constituted the chief mechanism of Lamarck's evolutionary theory and that Darwin rejected this mechanism.[21] We have shown the fallacy of both of these myths, but we do not suppose that we have seen the last of them.

THAT DARWIN WORKED ON HIS THEORY IN SECRET FOR TWENTY YEARS, HIS FEARS CAUSING HIM TO DELAY PUBLICATION

Robert J. Richards

There was no way in which Darwin was going to alienate important scientists by revealing his thinking on evolution—he did not do this for twenty years until he was forced into doing so.

—Michael Ruse, *Defining Darwin* (2009)

. . . what has become one of the major themes of this book: [is] Darwin's motives for his long delay in publication. His fear of persecution and ridicule was based not only on the unpopularity of evolutionary theory, but on the fiercer retribution meted out against proponents of materialism.

—Howard E. Gruber, preface to *Darwin on Man* (1974)

Two assumptions of long standing surround the history of Charles Darwin's (1809–1882) *On the Origin of Species:* first, that he worked on his theory for twenty years in secret and did so out of fear. But fear of what? A variety of fears have been attributed to him: fear of being charged with atheism, with materialism, or with bad science—of a charade of science that was embarrassingly speculative, comparable to that of his grandfather, Erasmus Darwin (1731–1802), or that of the French naturalist Jean-Baptiste de Lamarck (1744–1829) (see Myth 10). Then there is the second, associated assumption that these fears stayed his hand in publishing his theory, delaying the appearance of his great book.

Some time ago, I undertook an investigation of the way scholars dealt with Darwin's supposed delay in publishing the *Origin*.[1] I asked whether this was an interesting problem, for scholars, after all, do not wish to waste time on trivial issues but want to deal with truly interesting questions. But then, what makes an interesting question? Certainly a problem that touches on the principal ideas of a significant figure—that would be quite interesting—and Darwin's alleged delay in publishing meets this criterion. Another condition that determines the interest a problem might have is the expectations of the community of scholars. Darwin expended a huge amount of time and effort on his theory: from the period of his return from the *Beagle* voyage (1836) to the publication of the *Origin* (1859), he incessantly made entries related to his theory in his notebooks; he corresponded with a great many naturalists who might answer questions about species; he investigated nagging problems; he performed extensive experiments; and he began the composition of a book that would have dwarfed the *Origin*, which he regarded as the précis of this larger, never-published tome. So, Darwin's delay appears to be an interesting problem.

A final condition that would fix a problem as interesting is whether scholars have regarded it as such, meaning that when others have treated a question, subsequent scholars will often take it up. The scholarly concern with Darwin's delay began in the wake of the Darwinian anniversary celebrations in 1959, though the attention was scattered and unfocused. For instance, in his *Death of Adam* (1959), John Greene (1917–2008) mentioned in passing that Darwin was extremely cautious in advancing his "bold hypothesis," since his theory, as he admitted to his friend Joseph Hooker (1817–1911), was "like confessing a murder."[2] J. W. Burrow (1935–2009) thought Darwin hesitated because he feared his ideas might be mistaken for the thoroughly savaged transmutational views of Robert Chambers (1802–1871), whose *Vestiges*

of the Natural History of Creation appeared anonymously in
1844. Burrow urged that the dread of being taken for simply
another evolutionary speculator "haunted Darwin and enjoined
caution in announcing his views and patience in marshalling his
evidence."[3] Michael Ruse (b. 1940) concurred with Burrow: it
was anxiety of being taken for a fumbling amateur, like Mr. Vestiges, that caused him to falter.[4] Howard Gruber (1922–2005)
generalized what he took to be Darwin's primal fear. He scrutinized Darwin's notebooks and fell on certain passages that suggested the naturalist had become sensitive to his theory's materialistic implications, which Gruber dramatized as more destructive
of the traditions of Western civilization than evolution itself.[5]
Stephen Jay Gould (1941–2002) devoted an entire essay to the
problem of Darwin's delay; he endorsed Gruber's contention
that it was fear of the charge of materialism that shut Darwin
down.[6] And Adrian Desmond and James Moore, in their biography
Darwin: The Life of a Tormented Evolutionist (1991), found the
young Englishman's torment to lie in recognition that his theory's materialism aligned him with social radicals and could well
bring the opprobrium of his scientific peers crashing down on his
balding head.[7] Even the *New Yorker*'s pages have been breached
by the celebrity of the issue. Adam Gopnik begins his essay
(2006) on Darwin's accomplishment as follows: "Darwin's delay
is by now nearly as famous as Hamlet's," and offers what seems
the common view, namely, that Darwin delayed publishing
because he was "frightened about being attacked by the powerful and the bigoted."[8] Thus, from 1959 through the beginning
of the twenty-first century, scholars have offered quite a few reasons for Darwin's supposed two-decade delay in publishing his
theory.

But was there a delay? To suppose that Darwin delayed his
writing of the *Origin* suggests that the path was open for him to
have completed this task much earlier and that only some weak-
ness of resolve—perhaps an unwarranted fear—prevented him.
In my own essay, I made the commonplace observation that most

human actions are determined by a matrix of reasons that might impinge on an individual. I thought most of the causes suggested for the twenty-year interval had some weight, and that the task of the historian was to distribute this weight appropriately. But some important factors seemed neglected, namely the complexity of Darwin's growing account and his awareness of the large number of important problems he had to resolve for his theory to be successful.

One of those problems of considerable consequence was the phenomenon of the social insects: the "wonderful instincts" of worker bees and ants—the exact hexagonal cells of the honey bee, the slave-making behavior of some species of ants, and the self-sacrificing actions of soldiers among the social insects. In the 1840s, Darwin became quite worried over the apparent inability of natural selection to explain the cooperative and altruistic behaviors exhibited by these creatures, since selection enhanced the welfare only of the possessors of behavioral traits, not the recipients. But that wasn't even the most serious difficulty. Worker bees and ants are sterile; they leave no offspring to inherit any potentially beneficial behavior. In a manuscript of 1848, Darwin reckoned this "the greatest *special* difficulty I have met with."[9] And in the *Origin,* he stated flatly that he initially thought the problem of instincts of neuter insects "fatal to my whole theory."[10] This was not a difficulty he could let pass unnoticed, since it appeared that only divine wisdom could teach geometry to a honeybee. So here was a problem of significant proportions that did not easily yield a solution. Darwin resolved the difficulty only in 1858, in the throes of actually composing the manuscript that became the *Origin of Species.* Natural selection, he finally determined, operated on the whole hive or community to select just those insect groups that by accident had members displaying advantageous traits.[11]

I've now come to see two other significant problems that Darwin thought he had to solve before his theory could be unveiled. The first was the difficulty of divergence, a problem he neglected prior to the 1850s: What caused incipient species to diverge in character from one another and different genera to form even greater morphological gaps? He called it "the gravest objection which can be urged against my theory"—obviously forgetting he had nominated another perplexity for the principal source of ice in the blood.[12] Working out the problem of divergence caused him to add something like eighty manuscript pages to the composition of his book.[13]

Darwin knew that a respectable theory in the natural sciences should have a mathematical component, so he set out to mathematically demonstrate species formation. Using twelve large flora books, he statistically analyzed the number of large species against that of small species (that is, large species being those with a large number of varieties and small species being comparably determined); he also calculated the number of large genera (that is, with a large number of species) against that of small genera. His calculations seemed to confirm the pattern of species descent his theory predicted, namely that large genera had large species, which implied that species arose from earlier varieties. Because Joseph Dalton Hooker (1817–1911) recommended that the details of these calculations be omitted from the *Origin*—lest a tempting target be supplied to the mathematicians—Darwin discussed only the conclusions he derived and withheld the numbers.[14] He thus expended great amounts of time and labor in his botanical statistics and then suppressed the evidence.

In addition to solving major problems that took considerable time, there were the experiments Darwin performed to provide the empirical evidence that a good naturalist should marshal. He soaked seeds in seawater over weeks to determine sources of island vegetation; he raised fancy pigeons and crossbred them to uncover descent relationships; he planted plots of different grasses to compare competitive advantages; he dissected embryos of different species to show they more resembled each other than they did their adult forms. These kinds of experiments and others—all time consuming—provided the evidentiary base for his theory and gave Darwin the authority of an experimenter, showing that he was more than a passive observer.

These are just some of the major obstacles that Darwin had to overcome and the experiments he believed necessary to perform in order to present the most convincing argument for his theory. And, it must not be forgotten, he was making an argument, one "long argument" in his parlance.[15] His other major efforts in publication up to 1859 were mostly descriptive. When he did venture a fairly complex argument for the parallel roads of Glen Roy (1839), he got scorched by Louis Agassiz (1807–1873), who later showed that those Scottish ridges were formed not by marine action, as Darwin believed, but by retreating glacial lakes.[16] Darwin admitted his Glen Roy paper was "a great failure" and he was "ashamed" of it.[17] Crafting the complex argument of the *Origin* simply took time, especially as the possibility of devastating mistakes hovered over the enterprise.

In 2007, John van Wyhe contended that despite the asseverations and assumptions of many scholars over the last half century, the notion of a delay was simply a myth, as was the belief that Darwin kept his theory a secret prior to 1858.[18] Wyhe explained these assumptions as the result of scholars' having accepted uncritically the work of previous historians who didn't have access to the full Darwinian corpus of notebooks, manuscripts, and letters.[19] Wyhe focused his attention on the belief that fear held Darwin back from revealing his theory. He argued that there was simply no evidence that the stolid Englishman refrained from publishing because he blanched at the wrath of the elite.

It is quite difficult to plumb the mind of any individual to discover what motives might ground action or inaction, especially if that person has been dead for quite a long while. The case is a bit different, though, with Darwin. He did leave a broad trail of journal entries, essays, letters, and manuscripts. In light of such evidence, we can make the decently probable inferences that the two assumptions I have mentioned fall into the category of myths—remembering as we should, however, that there is often a degree of truth traveling in the guise of myth.

It is surely mythic to argue that fear froze Darwin's hand. He continued to work on his theory during the twenty-year interval in question, gathering evidence and unknotting the many difficult problems he faced, as well as dealing with a large family. The tormented evolutionist of Desmond and Moore's biography, an individual who cowered behind the arras, fearful either of the revolutionary mob or the scorn of men of high church and high pretension—that's a myth designed for dramatic effect. Yet Darwin did seem apprehensive about the "persecution of early Astronomers," and he did express his "fear [great evil] from vast opposition in opinion on all subjects of classification."[20] He frequently returned to the materialistic consequences of his theory.[21] He was, of course, quite cognizant of the scientific community's crushing dismissal of the evolutionary ideas of his grandfather, Lamarck, and Chambers. And we should recall it required a shove from Charles Lyell (1797–1875) to get Darwin started on his book. These accumulated signs, each slight in itself, do indicate some wariness, some restraint urging him to spend sufficient time making his theory as formidable as a British man-o'-war. The evidence suggests that Darwin was not an individual paralyzed by fear but one cautioned to make sure his construction could withstand the quakes of the intellectual world.

Wyhe contended that only scholars ignorant of the manuscript evidence would resort to the kinds of reasons just mentioned to explain an imaginary delay. But this argument collapses in a spring breeze. After all, it was precisely evidence from Darwin's notebooks that led Gruber to propose fear of materialism in the first place. Most serious scholars of Darwin's accomplishment are not innocent of the archive, and the signs derived therefrom suggest that reasonable trepidation might well have extended the stretch of time leading to the *Origin*.

Did Darwin keep his theory a secret before 1858? Wyhe asserts that scholars such as Greene, Loren Eiseley, Desmond and Moore, Peter J. Bowler, John Bowlby, Ruse, Janet Browne, Rebecca Stott, and David Quammen failed to mention that Darwin actually leaked his developing ideas to several close acquaintances. This can only be an incautious judgment on Wyhe's part. All of the aforementioned scholars certainly knew that Darwin revealed to Joseph Hooker in 1844 that his views were similar to Lamarck's and that his admission "was like confessing a murder."[22] Most of those scholars have recognized others of Darwin's circle to whom he made his theory known. Bowlby, for instance, lists quite a few such individuals: Charles Lyell, John Henslow (1796–1861), George Waterhouse (1810–1888), Joseph Hooker, Leonard Jenyns (1800–1893), Thomas Wollaston (1822–1878), and Asa Gray (1810–1888).[23] Ruse's judgment, mentioned in the epigraph to this essay, is fair: no prominent naturalists, such as William Whewell (1794–1866), Richard Owen (1804–1892), or Adam Sedgwick (1785–1873), did get advance word of Darwin's theory before 1858, when the joint essays of Alfred Russel Wallace (1823–1913) and Darwin were published in the *Journal of the Proceedings of the Linnean Society*.[24] So, we are left with a secret, but one that Darwin couldn't quite keep.

The assumptions that Darwin delayed publishing his theory and that he kept silent about his work are myths, legends, but ones that surround more than a bit of truth. And printing these legends does signal the great consequence Darwin's theory has had for contemporary intellectual and moral life.

THAT WALLACE'S AND DARWIN'S EXPLANATIONS OF EVOLUTION WERE VIRTUALLY THE SAME

Michael Ruse

Wallace set forth clearly the same theory as Darwin's, which he had conceived independently, that species of animals have evolved from each other through the action of natural selection.

—George Ledyard Stebbins, *Processes of Organic Evolution* (1971)

You said this when I explained to you here very briefly my views of "Natural Selection" depending on the Struggle for existence.—I never saw a more striking coincidence. If Wallace had my M.S. sketch written out in 1842 he could not have made a better short abstract. Even his terms now stand as Heads of my Chapters.

—C. R. Darwin to Charles Lyell (June 18, 1858)

Let's start with undeniable facts. Charles Darwin (1809–1882) and Alfred Russel Wallace (1823–1913) both became evolutionists; they both discovered natural selection, even though it was Darwin who alone used this term until Wallace urged on him Herbert Spencer's (1820–1903) alternative "survival of the fittest"; they both saw Thomas Robert Malthus (1766–1834) as a key influence in their discovery; and they both saw adaptation to circumstances as a crucial part of the process/product of evolution. Oh, yes—and the two made their discoveries independently.

There is a good reason why the names of Darwin and Wallace are linked. There is also a good reason why the revolution is "Darwinian" and not "Wallacean." Apart from the fact that Darwin got there twenty years before Wallace, from the very beginning (the *Sketch* of 1842) Darwin had a full theory in mind, covering a Whewellian "consilience of inductions" from social behavior, paleontology, biogeography, systematics, morphology, and embryology. Although Wallace was to do seminal work in such areas as biogeography, this was only done later. It is not there at the end of the 1850s, when Darwin and Wallace first published.[1]

This said, we have all known that there were differences, and I must say that rereading Wallace's 1858 essay (the one he sent to Darwin) for the first time in several years, I am struck by how great are the differences. I am not into counterfactual history with the enthusiasm of Peter Bowler, and I feel sure that today we would have something along the lines of the theory of evolution that we do have today—apart from anything else, it is true—but I see history with a Wallace and no Darwin as being a bit different, starting with the fact that Wallace might have had trouble getting his paper published and noticed.[2]

The most obvious difference between Darwin and Wallace—one that people universally comment on—is that whereas the domestic world and artificial selection are a fundamental part of Darwin's theory, not only are they not part of Wallace's theory but he goes to some considerable effort to deny their relevance. Wallace agrees with general opinion that domestic change is never permanent, and so he must show that it is unlike natural change. This he does by suggesting that domestic changes, like shorter legs or fatter bodies, would always be deleterious in the wild and so could never persist. Hence, domestic changes in some inherent way are different from natural changes.

In my youth, trained as a philosopher in the logical empiricist tradition, I would have argued happily that none of this is very significant. I would have argued that the domestic world was not really an essential part of Darwin's argument—which begins only with Malthus and the struggle for existence—and so the talk of the farmyard and the breeders' clubs could be sloughed off. Darwin and Wallace are back together again. Forty years of doing the history of science has convinced me that this will not do. The domestic world is in the *Origin* (and the earlier versions) just as much as embryology is, and sauce for the gosling (as one might say) is sauce for the goose and gander.

We know why the domestic world is in Darwin's theory.[3] Unlike everyone else, Darwin had a rural background and knew of the great advances that had been made by breeders in the previous half century or so. His uncle Josiah Wedgwood II (1769–1843)—the father of his wife, Emma (1808–1896)—was a gentleman farmer, much into trials with artificial selection. Darwin knew how effective induced change could be. Also, Darwin was (certainly compared to Wallace) far more methodologically sophisticated. Artificial selection gave him the handle to an empiricist *vera causa* backing for natural selection—a demand that figured greatly in the thinking of Darwin's intellectual mentor John F. W. Herschel (1792–1871).[4] Artificial selection was a force produced by human will that made plausible the analogous force of natural selection, something we cannot experience directly (or so Darwin thought).

I would today take matters further. Artificial selection is inherently bound up with the metaphor of design. Darwin quoted the breeder John Sebright (1767–1846) on this. You decide the kind of animal or plant you want and then you set about producing it. Likewise, natural selection is speaking to design—the functioning of adapted organisms, as picked up by the natural theologians, such as William Paley (1743–1805), and used as proof of the existence of God (see Myth 8). I am not now interested in whether Darwin was likewise theologically concerned (early he did think there was proof, later he went the other way); I am not now interested in whether my good friend Robert J. Richards is right in thinking this points to the neo-*Naturphilosoph* Darwin seeing life forces pervading nature (he didn't, and Richards is wrong); and I am not now interested in whether Stephen J. Gould (1941–2002) was right in thinking that this led Darwin into undue adaptationism (it didn't). I am concerned in stressing that you can't drop metaphors—I really have gone beyond logical empiricism!—and that this metaphor of selection leading to design is an inherent part of Darwin's theory in a way that it is not for Wallace.

It is true that Wallace (in his 1858 essay) acknowledges adaptation and its importance, and later (in the 1860s), under the influence of Darwin, he does important work on the adaptations of butterflies and moths; but in the essay I just don't see the design intoxication that we find in Darwin, from the first jottings to the *Origin* (and later). Wallace sees adaptation as important for success, but he is not using design to peer into the very nature of organisms, as does Darwin. To use Aristotelian categories, there is something deeply teleological about Darwin's thinking that I don't sense in Wallace. Final cause is also at issue. For Wallace, one bird flies better than another simply because it has stronger wings—there is something sadly ironic that his example of a success is that of the passenger pigeon—whereas for Darwin, the better wings exist to fly faster. In the end, of course—*pace* Richards—they are both mechanists (this is all before Wallace became a spiritualist), but at the very least, Darwin has a heuristic tool that Wallace does not have: a way to look for adaptation in a way that Wallace doesn't.[5]

So much for the domestic-world analogy; I'll come back to it at the end. I want to turn now to something that Peter Bowler spotted some thirty years ago.[6] Both Darwin and Wallace realize that the Malthusian struggle takes place among individuals. One bird flies faster than another and thus escapes the predator. For Darwin, this is fundamental. Change starts (and I would say ends) here. "Hence, as more individuals are produced than can possibly survive, there must in every case be a struggle for existence, either one individual with another of the same species, or with the individuals of distinct species, or with the physical conditions of life."[7] For Wallace, however, the individual struggle seems more of a purifying process within the group. If you are second rate, you are going to get wiped out. Change is a group phenomenon. One variety is going to do better than another. The very title of Wallace's essay flags the reader to this: "On the tendency of varieties to depart indefinitely from the original type." More than just this, the change doesn't seem to come about because one variety goes to war with another variety and wins—at least not directly. It is all a matter of change of circumstances and of one variety doing better in the new circumstances than do other varieties. So if you have a new variety within a species and circumstances change and the new variety does better than the mother species, the mother species gets eliminated and change has occurred. This is very different from Darwin's picture, in which change can certainly occur because of a change of circumstances—a new predator, for instance—but can also occur internal to the group, like when one form uses a little less food than the others.

In a way, therefore, I see Wallace as being more passive than the active Darwin. For Wallace, you wait for the world to impose change. For Darwin, you can start the change yourself. I am not sure if this has any linkup with Darwin's acceptance of the inheritance of acquired traits (so-called "Lamarckism") and Wallace's rejection thereof—another difference, and obviously here one that posterity judges Wallace right and Darwin wrong (see Myth 10). More significantly, I see this as a clear mark of another difference that always existed between Darwin and Wallace. Darwin, the child (particularly the grandchild) of industrialists, always bought into the Adam Smith (1723–1790) philosophy—that is, nobody does anything for anybody else without a hope of return. "It is not from the benevolence of the butcher, the brewer, or the baker that we expect our dinner, but from their regard to their own interest."[8] He was, in today's lingo, ever an individual selectionist. Wallace, the man who first heard the socialist Robert Owen (1771–1858) as a young teenager and who at the end of his life said that Owen was one of his greatest influences, was ever a group selectionist. For Darwin, it is always one organism against all others—allowing that the Darwin-Wedgwood-embedded Darwin (he married his own cousin, as did his sister, and remained cocooned at Down after he became sick) considered the family to be part of self. For Wallace, it was always one group sticking together and going through life pressured by the outside world. It is worth noting that Darwin and Wallace became very much aware of their differences on this matter and in the 1860s argued the topic fully without either budging the other.[9]

At this point I join together the difference over the domestic world and the difference over the focus of change, individual or group. From the beginning, Darwin had a secondary mechanism— sexual selection—and this he divided into male combat and female choice (see Myth 14). It is obviously an idea that comes from the domestic world, where we have selection for food and other useful attributes and selection for our sport and amusement, and where the latter generally comes in one of two forms: fighting animals (usually males) and beautiful animals (again often males). Up to and including the *Origin,* sexual selection was really secondary, but with Darwin, it became incredibly important. Shocked by Wallace's apostasy, arguing that humans could not have been created naturally, but agreeing that Wallace was right that many features (such as hairlessness) cannot be explained by natural selection, Darwin turned to sexual selection for support, and this mechanism became a major topic and tool in *The Descent of Man* (1871).[10]

Wallace expectedly had nothing on sexual selection in his 1858 essay, and then, although he accepted sexual selection through male combat, quickly rejected sexual selection through female choice, arguing that the reason why males are often gaudy and females rather drab has nothing to do with choice and everything to do with camouflage. Females sitting passively with their eggs or young needed to escape detection, and so they evolved colors accordingly.[11]

I need hardly say that sexual selection is the epitome of individual selection. It takes place only among members of the same species, and for sure no one is doing anything for anyone else. It is quintessentially Darwinian. I don't know that Wallace rejected sexual selection because it was individualistic—he always accepted male combat—but he had no urge to go that way, and the drab-females hypothesis does rather fit into his picture of change coming to varieties because of outside pressures.

Of course Darwin and Wallace had the same theory, but when you start to dig into things, beneath the surface their theories were not quite as similar as most people assume.

THAT DARWINIAN NATURAL SELECTION HAS BEEN "THE ONLY GAME IN TOWN"

Nicolaas Rupke

Charles Darwin . . . expressed the beauty of evolution in the famous final paragraph of the book that started it all—*On the Origin of Species* (1859): "from so simple a beginning endless forms most beautiful and most wonderful have been, and are being, evolved."

   —Jerry A. Coyne, *Why Evolution Is True* (2009)

We are surrounded by endless forms, most beautiful and wonderful, and it is no accident, but the direct consequence of evolution by natural selection—the only game in town, the greatest show on earth.

   —Richard Dawkins, *The Greatest Show on Earth: The Evidence for Evolution* (2009)

The year 2009 marked the 150th anniversary of the first edition of Charles Darwin's (1809–1882) *On the Origin of Species* (1859). Among the highlights of the celebrations were two best-selling exposés of evolutionary theory today, one by the University of Chicago biologist Jerry Coyne (b. 1949) and the other by Richard Dawkins (b. 1941), who was Oxford University's Professor for the Public Understanding of Science. Each author wrote a blurb praising the other man's book, unified as they are in claiming that Darwin's notion of evolution by natural selection is the only viable explanation for "endless forms most beautiful"—that is, the outstanding variety of living forms. Both men present much of their evidence for evolution as an argument for the efficacy of natural selection. Darwinian theory, Dawkins concludes, is "the only game in town." The sole serious challenge to Darwin has come from the creationist belief in intelligent design, not from science—or so they contend.

This claim is questionable and substantially incorrect—it is a myth—because a scientific alternative has been in existence from well before the publication of *The Origin of Species,* an alternative that does not invalidate natural selection altogether as a driving force of evolutionary change, but reduces it to a contributory factor in the origin of organic form and diversity on earth. My purpose in this essay is less to argue a scientific or a philosophical and theological point of view than to treat the matter historically.

There is nothing new about putting Darwin's views forward as the one and only scientific theory of the origin of species by presenting it as the alternative to the biblical belief in special creation. Contrasting Moses and Darwin—so to speak—was in fact initiated by Darwin himself when he added to the fourth edition of *The Origin of Species* "an historical sketch" of evolutionary thought. "Until recently," he wrote, "the great majority of naturalists believed that species were immutable productions, and had been separately created. . . . Some few naturalists, on the other hand, have believed that species undergo modification, and that the existing forms of life are the descendants by true generation of pre-existing forms."[1] These few naturalists, however, were said to be mere forerunners of Darwin, imperfectly grappling with evidence; and according to one admirer, Darwin provided "the magnificent synthesis of evidence, all known before, and of theory, adumbrated in every postulate by his forerunners—a synthesis so compelling in honesty and comprehensiveness that it forced men such as Thomas Henry Huxley [1825–1895] to say: How stupid not to have realized that before!"[2]

To repeat: the equating of evolution with Darwinian theory, as initiated by Darwin himself and perpetuated today by Coyne and Dawkins, is a myth, presented in the form of a historical narrative of evolutionary biology that in part defined itself in opposition to creationism while ignoring notions of structuralist (or formalist) evolutionary theory—that is, the attribution of the origin of life and of the many forms it has taken primarily to the operation of physico-chemical and mechanical forces, and less to natural selection, the impact of which, however, is not denied (more to follow).[3] The mythical part of the Darwinian account is based on several narrative ploys, such as forgetting, ignoring, and misrepresenting. One of the most blatant and enduring misrepresentations has been Darwin's allegation that his structuralist critic Richard Owen (1804–1892), founder of London's Natural History Museum, was a creationist.[4] A century and a half later, Dawkins continues to repeat this misrepresentation, in spite of the fact that historical scholarship has long restored Owen to his identity as an evolutionist, albeit not a Darwinian.[5] Other scientific critics of Darwin's theory, too, have been put under a cloud of suspicion for their alleged creationist or crypto-creationist solicitude.[6]

Let me briefly summarize some of the characteristic points and great names of the pre—and post–*Origin of Species* structuralist tradition. From the start—the late eighteenth century—the structuralist approach to the origin of species was a more comprehensive one than Darwinian theory ever was or is, if only because the question of the origin of organic diversity was treated as intrinsically related to the question of the origin of life—abiogenesis. This was believed to be a natural process—spontaneous generation—engendered by a constellation of material conditions and molecular forces. It connected the evolutionary history of life with the evolution of the earth, the solar system, our galaxy, and the elements. The literary epitome of this grand synthesis was Alexander von Humboldt's (1769–1859) *Cosmos: A Sketch of a Physical*

Description of the Universe (5 volumes, 1845–1862), whereas a more amateur and outspoken rendition was Robert Chambers's (1802–1871) *Vestiges of the Natural History of Creation* (1844).[7] They saw the history of the cosmos as an integrated process of material complexification that followed natural laws, and they understood the origin of life and species—organic evolution—as a process driven by molecular forces, "molecular evolution" in today's terminology. Owen referred to the formative forces as "inner tendencies." A century after Owen, Nobel Laureate Erwin Schrödinger (1887–1961), who made groundbreaking contributions to quantum mechanics, argued along similar structuralist lines in his *What Is Life? The Physical Aspect of the Living Cell* (1943), while today Simon Conway Morris speaks of "deep structure" and Keith Bennett writes "that macroevolution may, over the long-term, be driven largely by internally generated genetic change, not adaptation to a changing environment."[8]

During the early years of biology, around 1800, when the term "biology" was coined and the subject took institutional shape, many believed that each species, including *Homo sapiens,* had originated from a spontaneously generated germ, and that species were autochthonous (that is, they had originated naturally in the locations where they were found). Distinct global provinces of geographical distribution together with the phenomenon of sudden complexity in the paleontological record seemed to make sense that way.[9] Yet through the first half of the nineteenth century, a consensus developed that most species had come into being through a process of descent, although not gradually via an accumulation of small modifications but often instantaneously as the result of large changes, the latter exemplified by congenital conditions such as *situs inversus,* as in the case of dextrocardia (when the apex of the heart is oriented toward the right side of the chest, not the left side), or by dramatic metamorphic changes, like the ones known from metagenetic cycles (when sexually and asexually reproducing generations alternate).

Crystallography was brought to bear on organic form, its regularities, and its symmetries, highlighting striking similarities between crystals and skeletons (more to follow). Throughout a century of research along these lines, from the early nineteenth to the early twentieth centuries, several prominent scholars featured life's mathematical describability, such as Carl Gustav Carus (1789–1869), Ernst Haeckel (1834–1919), and D'Arcy Thompson (1860–1948), the last in his compendious *On Growth and Form* (1917).[10] Inner architectural logic of form was considered a more fundamental feature than the externalities of unpredictable form changes in response to environmental conditions. The molecular nature of the driving process explained the fact that not only crystals but also organic forms can be captured by arithmetic and geometry—Fibonacci sequence, golden ratio, crystal symmetry— as in the case of phyllotaxis, the likely biocrystalline nature of diatomaceous and radiolarian skeletons, and many instances of symmetry, especially in plants; later examples came from viruses, the helical structure of DNA, and the fractal nature of fern leaves, as well as the self-similarity in Ediacaran rangeomorphs. Life's forms express a structural logic and are, to a certain extent, predictable. The innumerable instances of convergent evolution are suggestive of patterns, of preferential pathways of evolutionary change that allow for notions of direction—perhaps orientation toward an end, a goal, as most recently suggested by Conway Morris in his *Life's Solution* (2003).[11]

The question has to be asked: How can this myth of Darwinism's singularity as a theory of evolution have taken hold? How can Darwin, in his Dawkinsian incarnation, have turned into the schoolyard bully of evolutionary biology? Here is not the place to go into the detailed scientific arguments over how much of organic diversity can be explained in structuralist terms and to what extent natural selection has acted as a morphing influence. Rather, I'd like to look at the issue of structuralism versus Darwinism in terms of the geography of knowledge and underline the importance of location. More fundamentally, in order to understand the victory of Darwinism over the older, more comprehensive structuralist approach—a victory that took place in the wake of World War II—we need briefly to step back from the scientific issues and look at the geography of evolutionary knowledge, at the locations where Darwinism flourished, at the sites where the structuralist approach took shape. David Livingstone has shown in his recent Gifford Lectures how differently Darwin was read, at times appropriated, in various places.[12] I'd like to extend this contention by arguing that nature itself was read differently in various locations and adjusted to the sensibilities of place and time. It is essential to realize that the two approaches to organic evolution were, by and large, products of distinct national cultures. As John C. Greene remarked some time ago:

> It is a curious fact that all, or nearly, all, of the men who propounded some idea of natural selection in the first half of the nineteenth century were British. Given the international character of science, it seems strange that nature should divulge one of her profoundest secrets only to inhabitants of Great Britain. Yet she did. The fact seems explicable only by assuming that British political economy, based on the idea of the survival of the fittest in the marketplace, and the British competitive ethos generally predisposed Britons to think in terms of competitive struggle in theorizing about plants and animals as well as man.[13]

Darwin scholarship of the past half century has informed us about the debt Darwin owed to British political economy, especially Thomas Malthus's (1766–1834) *An Essay on the Principle of Population* (six editions between 1798 and 1826); moreover, it has highlighted the considerable extent to which Darwin's thought was conditioned by the functionalism of the design argument and William Paley's (1743–1805) *Natural Theology* (1802). Darwin's theory of evolution was a product of British—more particularly, English—culture.[14] His theory of evolution by natural selection, with its emphasis on adaptation, was an anastomosis of Malthusian and Paleyan thought, and although Darwin inverted Paley's argument of design, one can argue that his theory remained an integral part of homegrown functionalist preoccupations. Dawkins is a present-day continuation of the same concern, engaging with modern-day proponents of intelligent design whose geographical and sociopolitical heartland continues to be the English-speaking world (in spite of the existence of differences between them).

By contrast, structuralism was primarily Continental, more specifically German. Most structuralists were Germans; Johann Friedrich Blumenbach (1752–1840) in Göttingen, and his students Lorenz Oken (1779–1851) and Gottfried Reinhold Treviranus (1776–1837) are just a few of the early influential names. Perhaps most prominent are Johann Wolfgang von Goethe (1749–1832), active in Jena and Weimar—the cultural heartland of the emerging nation-state of Germany—as well as the Dresden polymath and Goethe biographer Carus, author of the structuralist classic *Von den Ur-Theilen des Knochen- und Schalengerüste* (*On the Fundamental Components of Endo- and Exo-Skeletons*) (1828) which advanced the architectural approach to organic form in the context of German Romanticism and Idealist philosophy. Throughout the nineteenth century, Jena remained a hotbed of the morphological method, some major representatives of which were Karl Gegenbaur (1826–1903) and Ernst Haeckel.[15]

Haeckel died in 1919, just after the end of the Great War, in which Germany suffered humiliating defeat. Still, during Haeckel's lifetime, the German approach to organic evolution endured in the form of *Evolutionsmorphologie* and an interest in the mechanics of growth and form, *Entwicklungsmechanik*. However, German defeat in World War II seriously damaged structuralist evolutionary theory. During the Third Reich, academics and political ideologues appropriated German Romanticism and Idealism, in particular Goethe and Humboldt. A Nazi biography of Humboldt accentuated the latter's debt to Goethe's Idealism, and both men were presented as precursors of National Socialism.[16] Several structural evolutionists, such as the Austrian cofounder of paleobiology Othenio Abel (1875–1946), were outspoken anti-Semites and active participants in Nazi politics; the botanist Wilhelm Troll (1897–1978), known for his inflorescence studies, combined Idealist morphology in the tradition of Goethe with political proximity to Adolf Hitler (1889–1945). Through the period of the Third Reich, Darwinism became thought of as "un-German."

In the wake of World War II, putting forward a structuralist theory of organic evolution that was besmirched by having the fingerprints of Nazi collaborators and sympathizers all over it proved inopportune, to put it mildly. All the same, a number of senior scientists—among them Tübingen University mineralogist and Goethe scholar Wolf von Engelhardt (1910–2008)—made a cautious attempt to perpetuate and rehabilitate the structuralist tradition, but their success was limited, as the journal in which they published their views did not survive for long. Another Tübingen figure, the formidable champion of "a saltational, internally driven evolutionism in the Continental formalist tradition," Otto Schindewolf (1896–1971)—a man who, unlike Abel, had not blotted his political copy book during the Third Reich—was also unable to keep the younger generation of paleontologists and historians of biology, including Wolf-Ernst Reif (1945–2009), from turning to "the longstanding English preference for functionalist theories based on continuous adaptation to a changing external environment."[17]

Being a Darwinian in post–World War II Germany, and in neighboring countries that had collaborated, became something of a *Persilschein*—a certificate that helped cleanse a person of Nazi dirt and smudges, or simply a means of distancing oneself from the recent political past and joining cultural traditions of the victorious Allies. British Darwinians, in turn, were now able to denigrate structuralist thought on the basis of its German connotation. Today, Dawkins belittles the notion of "fundamental body plans," known by the German word *Baupläne*.[18]

In conclusion, perhaps one further point should be made. The success of the myth of the Darwinian evolution being "the only game in town" may in part be due to the fact that the theory of evolution by means of natural selection has enormously benefited from its many historical accounts, from Darwin's own early story to Ernst Mayr's *The Growth of Biological Thought* (1982),[19] and the hundreds of other articles and books that the "Darwin industry" since 1959—the centenary of *The Origin of Species*—has put into circulation. The formalist tradition in evolutionary biology has never received its historical exposition—that is, it has never been the subject of a comprehensive narrative. There is a need to recalibrate the historiography of evolutionary biology by writing a comprehensive account of the structuralist tradition.[20]

THAT AFTER DARWIN (1871), SEXUAL SELECTION WAS LARGELY IGNORED UNTIL ROBERT TRIVERS (1972) RESURRECTED THE THEORY

Erika Lorraine Milam

> Like almost everything else important in evolutionary theory, Charles Darwin discovered sexual selection. . . . After Darwin, the concept of sexual selection was largely ignored until 1972, when Robert L. Trivers resurrected the topic.
>
> —John Alcock, *Animal Behavior: An Evolutionary Approach* (1989)

It took time for Robert Trivers's (b. 1943) professional reputation to grow. Before enrolling as a biology graduate student at Harvard University in the late 1960s, Trivers worked for a small Cambridge-based educational organization helping to produce a yearlong program for elementary school children called Man: A Course of Study (or MACOS). He first learned about evolution within this context, crafting short books such as *Animal Adaptation, Natural Selection,* and *Innate and Learned Behavior.*[1] He had majored in history as an undergraduate, and his experience with MACOS gave him new hope for elaborating the reasons why humans (and animals) behave as they do. Thus inspired, Trivers published several papers in the early 1970s that became crucial to the burgeoning field of sexual selection over the course of the next decade. As increasing numbers of biologists focused their re-search on the mating behavior of animals, they cited Trivers's pa-pers—on reciprocal altruism, parental investment, the sex ratios of offspring, and parent-offspring conflict. As sexual selection found its footing as a field, so, too, did Trivers as one of its intellectual leaders. The suggestion, however, that Trivers single-handedly "res-urrected the topic" following a century of neglect could make sense only after redefining sexual selection as a concept studied solely in animals under natural conditions.[2]

Given his commitment to natural selection, Charles Darwin (1809–1882) needed to explain how members of the same spe-cies could differ dramatically in their appearance and behavior. If natural selection shaped traits that fostered survival, then surely all members of the population would come to look rather sim-ilar. Sexual selection affected traits that improved reproduction, however, and created room for the evolution of stable differences between the sexes.[3] Darwin included two elements in his original formulation of sexual selection.[4] The first, male-male competition, specified that just as members of a species vie for access to re-sources such as food or territory, males compete for access to fertile females. The second, female choice, laid the basis of this competition at the feet of females, who prefer to mate with some males rather than others.

For Darwin, the give-and-take of these two forces—male com-petition for mates and female choice of prospective suitors—explained why in some species males and females look so different from each other: competition between males led to war arma-ments, such as horns, whereas female choice accounted for male beauty. For humans, moreover, Darwin used sexual selection to elucidate the origin of the apparent physical differences that con-temporary racial scientists used to define ethnic groups.[5] Both mechanisms implied a mental awareness of self and others that natural historians were reluctant to attribute to animals, a problem accentuated by Darwin's use of anthropomorphic language to elaborate female choice. That human men and women make such judgments all the time remained uncontroversial and indeed became a crucial element of the flourishing eugenics movement in Great Britain as well as abroad.[6]

In later decades, biologists explored the effects of mate choice in lizards, fish, and especially the ubiquitous, quickly breeding fruit flies that came to embody research in experimental population genetics after the Second World War.[7] Rather than conceiving of sexual selection as a mechanism creating morphological differences between the sexes, however, these investigators typically suggested that mate choice would act to maintain reproductive isolation between species in evolutionary time. Many population geneticists and theorists came to question a clear distinction between sexual and natural selection. Additionally, they studiously avoided the anthropomorphic language that had contributed to the skepticism with which Darwinian female choice was initially greeted.

For biologists such as Lee Ehrman (b. 1935) and Claudine Petit (1920–2007), female fruit flies were notoriously particular in accepting suitors, mating only with males of the same species (by way of contrast, such discrimination did not characterize the mating behavior of male flies). By the 1960s, even mathematically inclined zoologists, such as Peter O'Donald (b. 1935) and John Maynard Smith (1920–2004), modeled the conditions under which sexual selection might have measurable effects on the mating success of individuals and whether it might serve as a mechanism for speciation without geographic isolation.[8]

Even if these investigations were far from common within the zoological community, Trivers knew of their existence and drew on them in composing his first forays into evolutionary theory. To see Trivers as Darwin's long-awaited heir, in other words, required redefining research on female choice in previous decades as something other than investigations into sexual selection—discrediting it either because of its association with eugenic theory or because it merely established the mathematical plausibility of sexual selection as an evolutionary mechanism but provided no empirical proof. None of this, however, explains the skyrocketing importance of Trivers himself to the identity of sexual selection as a field.

In the first of his seminal papers on evolutionary biology, Trivers described his theory of *reciprocal altruism* as an intuitive concept—"you scratch my back and I'll scratch yours"—that could form the basis of social cooperation in human (and animal) societies.[9] Cooperation among relatives made sense to Trivers because any effort individuals expended on their relatives, even at great personal cost, would help spread the genes they shared in the next generation. More difficult was the question of why non-relatives behaved altruistically toward one another, when individuals that cheated the system could gain all of the benefit at no cost. By linking present cooperation to future benefit (a kind of anthropological gift culture applied to evolutionary theory), Trivers noted that cheaters, if caught, would be excluded from such mutually beneficial networks of exchange.[10] He thus explained the evolution of cooperative social behavior as a balance between innate tendencies to behave kindly and to cheat, and held in check with a healthy dose of suspicion, all without recourse to a universal altruistic impulse.[11]

Turning his attention to mate choice, Trivers reasoned that males and females invested different amounts of resources in their offspring—whereas a male peacock might father numerous chicks with several females, expending little energy taking care of his young, a female peacock devoted far more time, attention, and resources to ensuring the survival of the few hatchlings she could produce in a season.[12] This difference in parental investment meant that a male's reproductive capacity would far outstrip any female's potential reproduction. Thus, we should expect females to be far pickier in their mate choices than males, and Darwin had been right to emphasize female choice and male competition as the dominant mechanisms of sexual selection. Trivers further put his finger on the pulse of the times: men and women, in his scenario, wanted different things out of marriage and life. These underlying evolutionary urges might then provide one basis for the conflict between the sexes in contemporary society.[13]

Browsing through the pages of iterative editions of John Alcock's (b. 1942) textbook on *Animal Behavior* reveals both the centrality of Trivers's research to the expanding field of sexual selection and Trivers's mythic role within that community.[14] Alcock published the first edition of his textbook in 1975, only six years after he had earned his PhD at Harvard, where—given their mutual interests in animal behavior—he likely knew Trivers. Certainly Alcock greatly admired the sociobiological theories that Edward O. Wilson (b. 1929) generated at Harvard in the years after he left and the theoretical heft his former colleagues derived from applying game theory and mathematical models to the evolutionary process.[15] *Animal Behavior* ultimately went through nine editions and became one of the most popular college textbooks on the subject.[16] Not until the second edition (1979), however, did Alcock mention Trivers's contributions as "exceptionally important."[17] In the next rewrite, he expanded this view, suggesting that Trivers's work on parental investment had been "extremely influential in the rebirth of interest in sexual selection."[18] By the late 1980s, Alcock invoked Darwin as the founder of all "important" aspects of evolutionary theory, including sexual selection, and a couple of paragraphs later ascribed to Trivers the theory's "resurrection." Even to Alcock, then, Trivers's key role as an evolutionary theorist had not been obvious from the outset.

Over the course of the 1980s, a new generation of biologists appropriated sexual selection as a theory explaining the evolution of reproductive behavior in animals and humans. Their field-based research on animal courtship provided an important foil for the even more remarkable growth of molecular biology and genetics. These zoologists saw the previous decades as lacking both careful observations of sexual behavior in wild species and a theoretical frame by which to interpret them. By demarcating their discipline as newly important within the biological sciences, they simultaneously redefined its history as characterized by a substantial eclipse in sexual selection from the time of Darwin to the 1970s. In this context, Trivers became a hero who had vanquished the ignorance of earlier generations.[19]

In subsequent editions of Alcock's textbook, however, Trivers's prominence receded. Alcock began by distancing Darwin from Trivers, inserting between them several pages that described the nuances of sexual selection for current research.[20] Then he removed Trivers entirely as a central figure uniting the field, opting instead to review his contributions in four separate places, each associated with one of his main papers.[21] Why? As the study of sexual selection gained institutional grounding, garnered professional authority, and expanded dramatically, Trivers's early studies no longer held the community together. Additionally, Trivers had not produced more groundbreaking work along the lines of his first publications, and he was plagued by personality conflicts with colleagues.[22] Given the widely acknowledged eclipse of his theory, Darwin took precedence as the sole iconic figure in the history of sexual selection.[23]

A large number of issues played a role in building the myth that sexual selection had been eclipsed for a century, primarily because of female choice. Ideologically, after the Second World War, biologists sought to distance themselves from theories associated with eugenics. As professional scientists turned their attention to animal behavior, anthropomorphism became associated with amateurs and armchair scientists. Shifting dynamics within professional biology made a difference, too. Researchers investigating sexual selection in the 1970s worked primarily in natural environments and self-consciously dissociated their work from earlier laboratory research.

Even so, scientific heroes in textbooks last only as long as they are useful pedagogically or disciplinarily. For a brief time Trivers's story fulfilled both functions. Parental investment provided an easy conceptual bridge to more recent work in the field, and Trivers's theories anchored a methodologically diverse community. In succeeding decades, research on sexual selection has branched out in many new directions.[24] Trivers, in turn, has once again picked up questions over which he first puzzled while working for MACOS: In a world where clear perceptions matter, how can we understand the value of deceit and self-deception in our lives?[25] Despite the increasing divergence of these legacies, biologists continue to demarcate sexual selection from other ways of exploring the natural world, treating it as a static set of tools developed by Darwin to explain differences between the sexes. As a result, the myth that sexual selection was virtually forgotten for almost a century persists, albeit without Trivers as its sole redeemer.

THAT LOUIS PASTEUR DISPROVED

SPONTANEOUS GENERATION ON THE BASIS

OF SCIENTIFIC OBJECTIVITY

Garland E. Allen

[Pasteur] did not permit his religious convictions to influence his scientific conclusions. His canon as a scientist, he firmly stated, was to stand aloof from religion, philosophy, atheism, materialism and spiritualism. . . . It was Pasteur's point that the material conditions of life had momentous implications for religion, but that these implications must not interfere with the objective interpretation of experimental results.

—Thomas S. Hall, *Ideas of Life and Matter* (1969)

There is no question that Louis Pasteur (1823–1895) was a towering figure in nineteenth-century science: trained as a chemist and crystallographer, Pasteur is credited with discoveries in fields as diverse as organic chemistry (determining the basis for chirality, or optical rotation of polarized light), the microbial basis of wine and beer brewing, immunology (the basis for vaccination), and the biology of the phylloxeran pests that were destroying the French vineyards in the nineteenth century. However, so much that has been written about Pasteur has taken on the dimension of myth. He has been revered as an almost superhuman scientist in the vein of Isaac Newton (see Myth 6). For example, it was said of him in 1887: "In France, one can be an anarchist, a communist or a nihilist, but not an anti-Pastorian [*sic*]. A simple question of science has been made into a question of patriotism."[1]

Nowhere is the mythologizing more apparent than in discussions of the famous controversy into which Pasteur entered between 1861 and 1864 with his countryman, the naturalist Félix Archimède Pouchet (1800–1872), on the highly controversial issue of spontaneous generation. Pasteur opposed spontaneous generation while Pouchet defended it, both putting forward experimental evidence to support their respective positions. The myth has centered on Pasteur as a model of the unbiased scientist who follows the truth wherever it may lead, with no preconceived ideas, no philosophical or political influences, and complete impartiality in the gathering and interpretation of data. The myth has two dimensions that cloud our understanding of the nature of science as a process: one is personal, the image of Pasteur as the supreme (and unrealistic) exemplar of objective scientific practice; the other is an exaggerated, idealized view of what has been called the "scientific method" (see Myth 26). Using the tools of the historian, scholars such as John Farley (b. 1936) and the late Gerald L. Geison (1943–2001) have enabled us to debunk the myth and use the case study itself to present a more accurate picture of science as a human, social process.

In an evening lecture in April 1864 at the Sorbonne, Pasteur summarized the experiments that he expected would deliver the final blow to any serious scientific account of the spontaneous generation of microbial life. To a packed auditorium that included such luminaries as Princess Mathilde Bonaparte (1820–1904), niece of Napoleon I, and the writers Alexandre Dumas (1802–1870) and George Sand (1804–1876), Pasteur admitted that the issue of spontaneous generation smacked of serious philosophical and religious questions—namely, those of materialism, evolution, and atheism. However, he went on to deny that any of these issues made any difference in his judgment of the experimental differences obtained over the past four years between himself and Pouchet:

Neither religion nor philosophy, nor atheism, nor materialism, nor spiritualism has any place here. I may even add: as a scientist I don't much care. It is a question of fact. I have approached it without preconceived idea, equally ready to declare—if experiment had imposed the view on me—that spontaneous generations exist as I am now persuaded that those who affirm them have a blindfold over their eyes.[2]

This is the stereotype that has characterized much of the popular representations of science over the past century or more. The debate between Pasteur and Pouchet is important to see in its true context as one in which, despite Pasteur's assertion, political, philosophical, and religious ideology played a major, if unacknowledged, role in designing, carrying out, and interpreting experimental results.

For background, it is important to recognize that France had been in a state of political, social, and economic turmoil since the February Revolution of 1848, the resignation of King Louis-Philippe, and the establishment of the short-lived Second Republic in December 1848, under the presidency of Prince Louis Napoleon Bonaparte (nephew of Napoleon I). A resurgence of Republicanism brought with it claims for greater liberalization and extension of the franchise, while simultaneously curtailing the powers of the Catholic Church and increasing those of the Legislative Assembly. Under these circumstances, church and state joined forces in the face of their common enemies: republicanism and revolution. Thus, by the time Napoleon initiated his coup d'état in December 1851, proclaiming the Second Empire (and himself emperor), the union of church and state had been greatly strengthened, and the new emperor relied heavily on the support of the church to maintain his power. The polarization between the church–state coalition and the more radical Republicans only deepened in the following decade. Republicans promoted science, materialism, and atheism as means of progress, and rejected the obscurantism associated with the Catholic Church. In this context, spontaneous generation became a focal point of controversy because of its apparent association with the formation of life without divine intervention. Pasteur, as a well-known scientist, devout Catholic, political conservative, and opponent of spontaneous generation, was thus well placed to present an "impartial" but necessary refutation of spontaneous generation.[3]

Although the experiments of Francesco Redi (1636–1697) had shown that one of the best-known examples of supposed spontaneous generation—the appearance of maggots in putrefying meat—could be rejected by preventing contact by flies with the meat, the issue had continued to surface in the later eighteenth and early nineteenth centuries, particularly in France in the 1820s and 1830s in the debates between anatomist and paleontologist Georges Cuvier (1769–1832) and his opponents Jean-Baptiste Lamarck (1744–1829) and Étienne Geoffroy Saint-Hilaire (1772–1844).[4] Both Lamarck and Geoffroy had endorsed the possibility of spontaneous generation, the process being especially integral to Lamarck's broader transformist, or evolutionary, views (see Myth 10). The same association of transformism, spontaneous generation, and materialism reemerged in the political atmosphere of the Second Empire in even starker relief with the 1862 publication of the French translation of Charles Darwin's (1809–1882) *On the Origin of Species* by Clémence Royer (1830–1902).[5] Royer voiced her support not only for a naturalistic theory of the origin of species but also for materialism, atheism, and republicanism. Further fueling the fears of Catholics and other conservatives, a movement known as "higher criticism" emerged in the 1860s, which aimed to treat biblical texts as historical documents free from any notions of revelation or supernaturalism.[6] The revival of spontaneous generation in this social and intellectual climate predisposed major figures of the scientific establishment, such as Pasteur, against any proponents of such heretical views.

Pasteur's work in the 1850s on the souring of wine confronting the French wine industry had convinced him of the ubiquity of microbes in the environment and of their likely implication in all fermentations, the spoiling of foods, and the spread of human disease. An important component of Pasteur's germ theory was his opposition to spontaneous generation. All bacteria, he claimed, came from the reproduction of previously existing bacteria and were not spontaneously generated from nonliving organic matter. Admitting to spontaneous generation would have seriously undermined the integrity of the germ theory.

Pouchet walked into this charged political and scientific arena in 1859 (the same year that Darwin's *Origin of Species* appeared) with the publication of his book *Hétérogénie; ou Traité de la Génération Spontanée (Heterogenesis; or, Treatise on Spontaneous Generation).*[7] He must have been well informed on the inflammatory nature of the topic because he devoted considerable space (the first 137 pages) to denying any materialistic or atheistic implications for his belief in spontaneous generation. Indeed, he argued that it was to the greater glory of God that the creation of life should be a continuing process. He denied that spontaneous generation could occur with purely nonorganic materials, even with the same elements involved. Despite these caveats and precautionary statements, Pouchet nevertheless found his work inevitably associated with materialism, atheism, and republicanism, thus attracting the opposition of both church and state.

Beginning in 1858, Pouchet had performed a simple set of experiments that appeared to support his belief in spontaneous generation. He set up a series of flasks of hay infusion heated to about 100°C so as to sterilize their contents. He then introduced into the flasks artificially produced air, or oxygen passed through mercury (as a kind of filter), sealed the flasks, and left them at room temperature. After a few days, they were seen under the microscope to be teeming with bacteria. To counter the argument that he had not boiled the hay infusions enough, he boiled them at 200°C and 300°C, even close to the point of carbonization, and still bacteria appeared.

When Pasteur read Pouchet's paper, he responded by pointing out that since bacteria exist everywhere around us, it was likely that Pouchet had either not fully sterilized the original broth or not plugged up the flasks quickly or carefully enough, thus allowing bacterial contamination of the liquid. This contention led Pouchet to repeat his experiments several times, always with the same results. Pasteur had to admit that Pouchet's experiments were the most meticulously of any on spontaneous generation carried out to date. But there was still one possible source of error: Pouchet's method of passing heated air through the mercury trough before admitting it to the flasks of sterile broth. In a simple experiment, Pasteur took a drop of mercury from the surface of the trough and introduced it into flasks containing a solution of sterilized yeast and sugar water. Within 24–36 hours, the flask was seen to be teeming with microbial life. Conversely, if the mercury were also sterilized before being introduced into the flasks, no microbial life appeared. The source of Pouchet's spontaneous generation was the contaminated mercury itself.

The two scientists proceeded to exchange comments and letters in scientific publications. So intense did the discussion become, and so important was the resolution seen to be, that in 1862 the French Academy of Sciences announced a contest, with a cash prize to be awarded by a jury of scientists to the best presentation. Pasteur and Pouchet were the main contenders.

Pasteur first demonstrated that boiling yeast and sugar water in a flask and then immediately sealing it by melting the glass at the top so that the contents were not allowed contact with air prevented the growth of bacteria and hence decay of the broth. However, Pouchet countered with a perfectly reasonable argument: he pointed out that boiling might have changed the chemical composition of the air inside the flask, rendering it unsuitable for spontaneous generation. Pasteur then designed an elaborate apparatus that would answer this objection while supporting his contention that all supposedly spontaneous generations were the result of contact with microbe-laden particles from atmospheric air. He attached a flask of sterile broth to a glass tube with two stopcocks leading to a suction pump. By means of the suction pump, he passed air through a red-hot tube in a furnace and introduced it into a flask of sterile sugared yeast water, then sealed off the neck of the flask with heat. These flasks sat for months and showed no microbial growth. He then repeated this basic procedure: using normal (unheated) atmospheric air passed through a sterile wad of guncotton placed in the tube, he got the same results: no microbial growth. However, when he then allowed the piece of guncotton to be drawn into the flask, after a few days it teemed with microbes.

To counter any complaint that the guncotton filter may have altered the air passed through it, Pasteur performed another simple but elegant set of experiments with specially designed swan-necked flasks that allowed untreated atmospheric air to diffuse back and forth between the inside and outside with no filtration. He introduced sugared yeast water into these flasks and boiled them to kill any microorganisms and then let them sit for various periods of time. The reasoning behind the S-shaped design for the flask necks was that the lower portion of the neck would serve as a trap for the heavier dust particles and bacteria carried in the air that diffused in and out. Using this apparatus, Pasteur reasoned, would allow air to come in contact with the broth but no airborne bacteria would make it beyond the trap. If spontaneous generation could occur, then it ought to do so under these circumstances. The results of Pasteur's experiment were quite dramatic: even after several months there was no decay in the flask. Moreover, he made a bold prediction: if he tilted the flask so that some of the sugared yeast water got into the trap and was then returned to the main receptacle (Figure 15.1), the liquid in the receptacle should show microbial growth. When he carried out this experiment, bacteria appeared in the broth in just a few days. The French Academy of Sciences Committee awarded the prize to Pasteur.

In the course of the debate, Pasteur *seems* to have been driven by the sheer force of logic and ingenious experimental design. But the story is not quite as simple as it might at first appear. It is certainly the case that Pasteur's experimental work was exemplary. Yet Pasteur himself enjoyed the French government's financial support, was a loyal supporter of Napoleon III, and, on several occasions, was an invited guest of the emperor at one of his country estates. Pasteur was also a member of the French Academy of Sciences, at the time composed of the most pro-government, scientific elite of France. Pouchet, living in Rouen, was only a corresponding member and was not among the inner circle of elites that included Pasteur. On the religious side, Pasteur was a devout Roman Catholic who, as his Sorbonne speech in 1864 explicitly noted, understood all too well the relationship between the theory of spontaneous generation and liberal, materialistic, and "atheistic" ideas. This is important, as virtually all of the committee's members were strong Catholics and anti-Republicans who recognized that public affirmation of spontaneous generation was politically sensitive. Indeed, Pouchet, recognizing how stacked the committee was against his views, withdrew from the contest in 1862, only to revive his participation a year later at the urging of friends and associates.[8]

Perhaps the greatest irony in this story is that both Pouchet and Pasteur were right. The two men were using different sources of organic broth: Pasteur's was sugared yeast water, while Pouchet's was a hay infusion, a liquid prepared by soaking hay in water. Because he did not repeat Pouchet's experiments using hay, Pasteur did not discover that the natural bacteria found in hay infusions include some species that can form spores, which enable them to survive severe conditions, such as drought, cold, or heat. The spores present in Pouchet's preparation were heat resistant enough to survive the short boiling time and even the higher temperatures to which he subjected them. They were thus able to emerge from their dormant stage and start reproducing once the flask cooled down.

Several major conclusions come out of the analysis of Pasteur's work on spontaneous generation. One is that Pasteur used the image of the detached scientist to further solidify his position

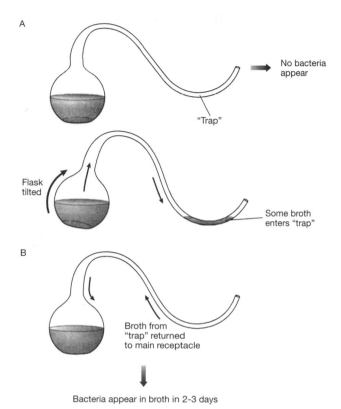

Figure 15.1. Diagram showing how tipping the swan-necked flask, so that unspoiled broth was brought into contact with the curved "dust trap" and then returned to the main receptacle, brought about the appearance of bacterial infestation. (Source: Allen and Baker, *Biology: Scientific Process,* 63.)

within the French academic establishment. The germ theory of disease had become an important component of both French colonial medicine and the campaigns for public health at home and, thus, among other things, a source of financial and political support for Pasteur and his Institut Pasteur.[9] A second conclusion from this study is that there is no such thing as the idealized

"scientific method" to which Pasteur and others of the late nineteenth and early twentieth centuries have appealed. This view,
which we have all seen promoted in textbook after textbook, has
little to do with the way science is practiced in the laboratory on
a day-to-day level. One of the most important outcomes of current work in the history, philosophy, and sociology of science is
that it demonstrates that science is a very human activity, bound
up with human ideals and aspirations.

THAT GREGOR MENDEL WAS A LONELY PIONEER OF GENETICS, BEING AHEAD OF HIS TIME

Kostas Kampourakis

By experimenting with pea plant breeding, Mendel developed three principles of inheritance that described the transmission of genetic traits, before anyone knew genes existed. Mendel's insight greatly expanded the understanding of genetic inheritance. . . . Mendel was curious about how traits were transferred from one generation to the next, so he set out to understand the principles of heredity in the mid-1860s. . . . Mendel . . . hypothesized that each parent contributes some particulate matter to the offspring. He called this heritable substance "elementen." . . . Indeed, for each of the traits he examined, Mendel focused on how the elementen that determined that trait was distributed among progeny.

—Ilona Miko, "Gregor Mendel and the Principles of Inheritance" (2008)

Before Gregor Mendel, theories for a hereditary mechanism were based largely on logic and speculation, not on experimentation. In his monastery garden, Mendel carried out a large number of cross-pollination experiments between variants of the garden pea. . . . From the precise mathematical 3:1 ratio . . . he deduced not only the existence of discrete hereditary units (genes) but also that the units were present in pairs in the pea plant and that the pairs separated during gamete formation. . . . Mendel also analyzed pure lines that differed in pairs of characters. . . . The F2 generation produced by self-pollination of F1 plants showed a ratio of 9:3:3:1. . . . From this result and others like it, he

deduced the independent assortment of separate gene pairs at
gamete formation.

—A. M. Winchester, "The Work of Mendel," *Encyclopedia
Britannica*

Gregor Mendel (1822–1884) is often considered to have been the
first and only scientist to study heredity during the nineteenth century. In addition, he is often presented as the founder of genetics
who understood that inheritance was particulate in nature, who
was ignored by his contemporaries, and whose reputation was
established posthumously, in 1900, with the rediscovery of his pioneering paper that contained the "laws" of heredity. Although
this account of Mendel's life and work has been questioned since
at least 1979, it still appears in biology textbooks and in other
widely read works. In this chapter I explain why this account is
not historically accurate and why it distorts how science is actually done.[1]

The historical evidence suggests that Mendel intended to study
hybridization in particular and not heredity in general; his 1866
paper, titled "Versuche über Pflanzen-Hybriden" ("Experiments
on Plant Hybrids"), does not indicate that he was thinking in
terms of hereditary particles; and the laws of segregation and of
independent assortment, which are attributed to him, certainly do
not appear in his paper in the way biology textbooks currently
describe them. Mendel is often presented as a lonely pioneer
whose work was ignored during his lifetime because it was ahead
of its time. However, questions about heredity emerged in a variety of cultural domains, and Mendel's work should be considered in its wider historical context.[2]

Mendel was born in Moravia, at that time a province of the
Austro-Hungarian Empire, which is now part of the Czech Republic. In 1843, he entered the Augustinian Monastery of
St. Thomas at Brno (Brünn), which at least since 1840 had been
a center of efforts to practice breeding under a scientific perspec-

tive. The abbot, Cyril Napp (1792–1867), supported Mendel's studies at Vienna, in physics with Christian Doppler (1803–1853) and in chemistry, paleontology, and plant physiology with Franz Unger (1800–1870). In 1853, Mendel returned to Brno and began his experiments, progressively becoming interested in the question of whether hybridization could produce new species. He was aware of the work of earlier hybridists, such as Josef Kölreuter (1733–1806) and Carl Friedrich von Gärtner (1772–1850), who had both come to the conclusion that hybridization could not produce new species. Unger, in contrast, had argued that this was possible. In fact, in his 1852 *Botanical Letters,* he rejected the idea of species fixity and argued that the plant world had developed gradually.[3]

This is perhaps what aroused Mendel's interest and prompted him to perform his hybridization experiments, starting during the summer of 1856. For his experiments he selected several varieties of the edible pea *(Pisum sativum)*. From these he obtained about thirty-four distinct varieties, which he subjected to a two-year trial for purity. Mendel focused especially on the inheritance of seven characters: the form of the ripe seed, the color of the endosperm, the color of the seed coat, the form of the ripe pods, the color of the unripe pods, the position of the flowers, and the length of the stem. He concluded his experiments with *Pisum* in 1863 and subsequently tried to confirm his conclusions with several species of the genus *Phaseolus*. In some cases, he obtained in the offspring the characteristic 3:1 ratio between the two parental characters, whereas in others, he did not.[4]

Mendel's results with *Pisum* were included in a paper that he presented at the meetings of the Brno Natural Science Society on February 8 and March 8, 1865, and that was published in the society's journal in 1866. In the beginning of the paper, Mendel expressed his aim "to follow the development of hybrids in their descendants." He also noted that until that time, "a generally valid law describing the formation and development of hybrids [had] not yet been established." [5] He observed that the hybrids obtained from the various crosses between different varieties were not always intermediate between the parental forms. He also noticed that there was often a close resemblance between a parental character and a hybrid one, calling such characters dominant. He called the other characters, which did not appear in the hybrids but that reappeared fully formed in the next generation, recessive. Mendel identified this generation, consisting of the offspring produced when hybrids were crossed, as the first generation, whereas in contemporary genetic parlance it would be the second one—F_2. In this chapter, I follow Mendel's terminology and refer to the generation bred from hybrids as the first one.

When Mendel calculated the number of offspring in the first generation of all the experiments, he found a roughly constant average ratio of 3:1 between the number of plants with the dominant characters and those with the recessive ones. Mendel observed that when the plants of the first generation exhibiting the recessive character (denoted as "a") bred, their offspring did not vary further. He also noticed that when the plants of the first generation exhibiting the dominant character bred, their offspring either exhibited the dominant and recessive characters in the proportion of 3:1 or exhibited only the dominant character. Mendel concluded that in the former case, plants possessed both a dominant and a recessive character (Aa), whereas plants in the latter case possessed a dominant character (denoted as "A"): "It is thus proven that, of those forms that possess the dominant trait in the first generation, two parts carry the hybrid character whereas one remains constant with the dominant trait." He also deduced that the 3:1 ratio resolved itself in all experiments into the ratio of 2:1:1. At this point, Mendel made the distinction between hybrid and ancestral characters or traits (using the terms "character" and "trait" synonymously). The dominant character could be either an ancestral one, which remained constant through generations, or a hybrid one, which segregated in the offspring. It should be noted that in the latter case Mendel did not refer to the traits of hybrids but rather to traits that were themselves hybrids.[6]

Given our current knowledge of genetics, it is easy to read too much into Mendel's paper. It is important to note that Mendel described the transmission of phenotypes (characters or traits) rather than that of hereditary elements or particles (like genes), as is clear from passages early in his paper:

> In the discussion that follows, those characteristics which pass into the hybrid combination completely or almost unchanged, and hence represent the hybrid traits themselves, will be described as "dominant," and those which become latent as "recessive." The expression "recessive" was chosen because the traits thus designated recede or disappear completely on the hybrids, but then reappear again unchanged, as will be shown later on, among the descendants of these same hybrids.[7]

As already mentioned, Mendel called the characters and traits—not any invisible, hereditary particles—as dominant or recessive. He used the word *factoren* (factors) in his paper, but it seems that he followed Gärtner in using this term to refer to the species hybridized rather than to any underlying factors influencing particular traits.[8]

Mendel went on to describe the offspring of hybrids that exhibited different characters, presenting the results of a cross between two plants that differed in two characters. He obtained 556 seeds from 15 plants: 315 exhibited the dominant characters, 101 exhibited one dominant and one recessive character, 108 exhibited the other recessive and the other dominant character, and 32 exhibited the two recessive characters. Mendel did not, however, describe this as the classic 9:3:3:1 ratio, found in much of the literature nowadays, but as a 1:1:1:1:2:2:2:2:4 ratio.[9] Although passages in Mendel's paper hint at the laws of segregation and of independent assortment, they never appear as "the laws." The section titled "The Second Generation of Hybrids" ends with the following conclusion:

> The ratio of 3:1 according to which the distribution of the dominant and recessive character occurs in the first generation, thus resolves

itself into the ratio 2:1:1 in all trials, if one all along distinguishes the dominant trait in its meaning as a hybrid-trait and ancestral character. Since the members of the first generation emerge immediately from the seeds of the hybrids, *it now becomes obvious that hybrids of two different traits respectively form seeds of which one half develops again the hybrid-form while the other gives rise to plants which remain constant and receive the dominant and recessive character in equal parts.*[10]

These words can be read as a description of segregation of alleles (dominant or recessive), which existed together in the hybrids and were passed to different offspring. Mendel described this as $A+2Aa+a$, but "A" and "a" denoted the parental characters, and "Aa" the hybrid characters, not some pairs of hereditary elements. This is different from the current description of segregation in textbooks, according to which when two heterozygotes (individuals possessing two different alleles, or forms of the same gene) are crossed ($Aa \times Aa$), there is a genotypic ratio of $1AA:2Aa:1aa$ in the offspring.

Later, toward the end of the section entitled "The Descendants of Those Hybrids in Which Several Differing Traits Are Joined," Mendel wrote the following:

There is therefore no doubt that the following statement is valid for all traits that were admitted to the experiments: *The descendants of the hybrids, in which several essentially different traits are conjoined, represent the members of a combinatorial series, in which the developmental series for two differing traits respectively are combined. At the same time this demonstrates, that the respective behaviour of two differing traits in hybrid association is independent of any other differences in the two ancestral plants.*[11]

These words brings independent assortment of alleles to mind. However, Mendel's own conclusions are different from what one would expect if one thought in terms of contemporary genetics.

Strictly speaking, Mendel did not discover the two laws commonly attributed to him; nonetheless, because of his experimental design, he perhaps observed their consequences under the particular experimental conditions. Neither did he discover that inheritance is particulate in nature; indeed, he does not clearly write about hereditary particles in his paper. Rather, as is evident in the previous quotations, Mendel investigated how particular characters developed in hybrids and their progeny. Mendel's main conclusion is that all sperm and eggs contain a single version of each of the characters they carry. When a sperm meets an egg containing a different version of a particular character, the resulting offspring will eventually contain both versions, which are independently transmitted to the sperm and eggs that the offspring will later produce.

In all cases, Mendel refers to hybrids; he does not seem to be developing a general theory of heredity. The mechanism of heredity was, however, at the center of biological thought during Mendel's time, in part because Charles Darwin's (1809–1882) theory of descent with modification through natural selection (published in 1859 in the *Origin of Species*) lacked a complementary theory that could explain the origin and inheritance of the variations so central to it. In response to this problem, various scholars—such as Charles Darwin himself, Herbert Spencer (1820–1903), Francis Galton (1822–1911), William Keith Brooks (1848–1908), Carl Wilhelm von Nägeli (1817–1891), August Weismann (1834–1914), and Hugo de Vries (1848–1935)—developed theories of heredity from an evolutionary perspective during the nineteenth century. All these people were influenced by one another in various ways, as can be seen in their references to the published works of the others. Mendel remained an outsider to this group; because his aims differed from theirs, it should come as no surprise that the German words for "heredity" or "inheritance" do not appear in his paper.[12]

Among the scholars in the aforementioned group, only Nägeli came to know of Mendel's experimental work; indeed, the two men corresponded from 1866 until 1873. Following Nägeli's advice, Mendel worked on *Hieracium* (hawkweed, a genus of the sunflower family) from 1866 to 1871 but failed to find an agreement between his results and conclusions with this plant and those with *Pisum*. This does not mean that Mendel's work remained in oblivion, because on at least one occasion Nägeli cited Mendel's 1866 paper. The Brno Natural Science Society also sent more than one hundred copies of the journal that included Mendel's paper to scientific centers around the world. At least ten references to Mendel's paper appeared in the scientific literature before 1900, some of them in books that were widely read by naturalists. Therefore, it was possible for Mendel's work to become more widely known. Yet it did not.[13]

In 1900, Hugo de Vries (1848–1935), Carl Correns (1864–1933), and Erich von Tschermak (1871–1962) published the results of their research on plant hybridization, which, as they noted, agreed with those obtained by Mendel. De Vries actually published two papers on this topic, but in the first one he did not mention Mendel. Correns, a student of Nägeli's who was probably long aware of Mendel's work, insisted on Mendel's priority over both de Vries and himself, perhaps in an attempt to resolve a potential priority dispute. This brought Mendel back into the scene.[14]

Mendel's paper quickly became the foundational document of the new science of genetics. However, it should be noted that the rapid acceptance of Mendel's work after 1900 was the result of a new conceptual framework, not simply of the discovery of new empirical data. Francis Galton and August Weismann had created a framework that supported the idea of "hard" heredity, with discontinuous variation and nonblending characters. In addition, cytologists had provided evidence supporting the view that particles inside the cells might be related to the emergence of characters. By that time, important advancements in cytology had taken place with the employment of vital dyes that allowed cellular structures, such as nuclei and chromosomes, and processes, such as meiosis and mitosis, to be visualized with new and improved microscopes. Thus, in 1900, Mendel's paper brought together the conclusions of breeding experiments with those of cytology, showing that particulate determinants, which existed in the nucleus of the cells, were segregated and independently assorted. Only in hindsight did Mendel become a pioneer of genetics.[15]

It is very interesting, though not widely known, that Walter Frank Raphael Weldon (1860–1906) showed in 1902 that Mendel's "laws of inheritance" might not actually work, even for peas. Weldon's studies of varieties of pea hybrids led him to conclude that there was a continuum of colors from greenish yellow to yellowish green between the green and yellow ones, as well as a continuum of shapes from smooth to wrinkled in gradually increasing degrees. It thus appeared that in obtaining purebred plants for his experiments, Mendel actually eliminated all natural variation in peas.[16]

Two main conclusions can be drawn from all the above. First, the usual presentation of Mendel as a heroic, lonely pioneer who discovered the principles of genetics and established the appropriate experimental approaches for the study of the phenomena of heredity not only distorts the actual history of genetics but also distorts how science in general is actually done. Science is a human activity, done within scientific communities, in particular (social, cultural, religious, political) contexts, and is rarely—if ever—pursued by isolated individuals. Second, Mendel carried out his experiments in the context of a series of practical questions related to agriculture and the socioeconomic context of Brno. Scientific questions frequently arise out of existing economic or technological ones, rather than just strictly theoretical considerations and curiosity. Mendel was trying to provide answers to practical questions, and he was an important and well-known contributor to the study of plant hybridization.

Therefore, contrary to what many textbooks still claim, Mendel "discovered" neither the particulate nature nor the general "laws" of heredity. He contributed virtually nothing to the development of a theory of heredity during the latter half of the nineteenth century, but when his paper was read in a new context after 1900, it became the foundational document of genetics. This does not undermine the importance of Mendel's experiments. But one should clearly differentiate between their impact in the context in which Mendel conducted them in the mid-nineteenth century and their impact in the new context in which they were eventually adopted as the foundations of genetics in the beginning of the twentieth century.

THAT SOCIAL DARWINISM HAS HAD A PROFOUND INFLUENCE ON SOCIAL THOUGHT AND POLICY, ESPECIALLY IN THE UNITED STATES OF AMERICA

Ronald L. Numbers

Social Darwinism was an influential social philosophy in some circles through the late 19th and early 20th centuries, when it was used as a rationalization for racism, colonialism, and social stratification.

—John P. Rafferty, ed., *New Thinking about Evolution* (2011)

Based on the work of Charles Darwin in explaining evolution, Social Darwinism is the application of "survival of the fittest" to society and business. It held that weaker (or unfit) companies would die off at the hands of stronger, better and superior corporations. It was seen as justification of laissez-faire capitalism, as government should not interfere in the natural economic evolution by which smaller, weaker companies are eliminated. This concept was also applied to individual success, by which the most successful in society were thought to be the smartest, hardest working and as such most fit.

—Preparatory Information for the New York State Regents Examination (1999–2011)

Despite all the talk about social Darwinism in the twentieth century, nineteenth-century Americans paid relatively little attention to the social implications of Darwinism. This was true of Charles

Darwin (1809–1882) himself, who devoted an entire book to *The Descent of Man, and Selection in Relation to Sex* (1871) but only occasionally prescribed human behavior based on his theory of natural selection. In a section of *The Descent of Man* headed "Natural Selection as Affecting Civilized Nations," he contrasted savage and civilized societies:

> With savages, the weak in body or mind are soon eliminated; and those that survive commonly exhibit a vigorous state of health. We civilized men, on the other hand, do our utmost to check the process of elimination; we build asylums for the imbecile, the maimed, and the sick; we institute poor-laws; and our medical men exert their utmost skill to save the life of everyone to the last moment. There is reason to believe that vaccination has preserved thousands, who from a weak constitution would formerly have succumbed to small-pox. Thus the weak members of civilized societies propagate their kind.[1]

In spite of the negative biological effects of many humanitarian efforts, Darwin generally defended them in the name of human nobility.[2]

Most American scientists either remained silent on the topic or denounced attempts to apply Darwinian principles to human society. The California naturalist Joseph LeConte (1823–1901), perhaps the most influential scientific popularizer of evolution in the United States, insisted that "natural selection will never be applied by man to himself as it is by Nature to organisms. His spiritual nature forbids." Even the geologist-anthropologist John Wesley Powell (1834–1902), who rejected traditional religion, nevertheless condemned efforts to apply "the methods of biotic evolution" to humans: "Should the philosophy of [Herbert] Spencer, which confounds man with the brute and denies the efficacy of human endeavor, become the philosophy of the twentieth century," he warned, "it would cover civilization with a pall and culture would again stagnate."[3]

The American disciples of the British philosopher Herbert Spencer (1820–1903)—not the Darwinians—led the effort to apply evolution to human societies. Claiming "the right to ignore the state," Spencer opposed government intervention in regulating commerce, supporting religion, educating the young, and even caring for the sick and poor—all in the name of science and ethics. "Unpitifying as it looks," he insisted, "it is best to let the foolish man suffer the appointed penalty of his foolishness."[4] Because his books sold impressively in the United States, it is easy to over-estimate his influence.[5] The biologist Henry Fairfield Osborn (1857–1935), marveling at the philosopher's popularity, noted that Spencer stated "the very A, B, C of science in language which is so obscure to a college professor that it must inspire awe among some thousands of his readers." It surely did not help Spencer's reputation in a predominately Christian country that he notoriously reduced divinity to the "Unknowable," a concept that few but skeptics and Unitarians could love.[6] Although the British philosopher came to be seen as an apologist for capitalism and unrestrained struggle for survival, his most recent biographer has convincingly argued that he has been misunderstood: "Spencer did not accept that modern individuals and societies would continue to make progress through struggle for survival."[7]

Despite the damnation of the orthodox, Spencer recruited several well-placed American apostles, including William Graham Sumner (1840–1910), Yale's celebrated professor of political and social science. Sumner, often portrayed as the country's foremost social Darwinist, did for a time teach his students that "if we do not like the survival of the fittest, we have only one possible alternative, and that is the survival of the unfittest."[8] But biological evolution remained marginal to his study of human societies, and by 1884 he had quit talking about the fit and unfit. He roundly condemned imperialism, especially around the time of the Spanish-American War.[9]

Another of Spencer's American disciples was Andrew Carnegie (1835–1919), the Scottish-born steel magnate and philanthropist who at times justified the concentration of wealth in terms of Spencerian evolution. Overall, however, neither Darwin nor

Spencer had much influence on American businessmen. As the historian Irvin G. Wyllie showed in an elegant study decades ago, "Gilded Age businessmen were not sufficiently bookish, or sufficiently well educated, to keep up with the changing world of ideas."[10] For ethical guidance they turned to the Bible, not Darwin's *Descent of Man* or Spencer's *Social Statics*. In one of the few references to biological evolution coming from the business community, the Ivy League–educated philanthropist John D. Rockefeller Jr. (1874–1960)—not his father, as is sometimes alleged—described the growth of a large business as "merely a survival of the fittest. . . . The American beauty rose can be produced . . . only by sacrificing the early buds which grow up around it."[11]

As we have seen, Darwin himself noted the dysgenic role of modern medicine. Surprisingly, few commentators took note. Like other educated Americans, physicians discussed the strengths and weaknesses of Darwinism but rarely its implications for public health. One of the few medical men to address the apparent conflict between evolutionary science and medicine was Charles V. Chapin (1856–1941), a national leader in the field of public health. In contrast to most of his colleagues, who simply ignored the alleged complaints of some biologists and sociologists that physicians were hindering "the beneficent workings of natural law by preserving those who are least fitted to survive, and permitting them to increase their kind," Chapin squarely faced the charge.[12] He admitted that preventive measures to control against infectious diseases interfered "with the action of natural selection by the artificial preservation of the unfit," but concluded, nevertheless, that physicians were not "warranted in abandoning our fight against these diseases and permitting natural selection to have again a free field."[13] Besides, preventing epidemics protected the strong as well as the weak.

Inattention to Darwin's provocative passage about how modern medicine thwarted the effects of natural selection in civilized societies gave way to heated discussion in the early 1920s, when the fundamentalist leader William Jennings Bryan (1860–1925) featured Darwin's statement as confirmation of his own negative view of evolution. In a widely circulated attack on "The Menace of Darwinism," Bryan dissected Darwin's "confession." The British naturalist, he alleged, "condemns 'civilized men' for prolonging the life of the weak. . . . Medicine is one of the greatest of the sciences and its chief object is to save life and strengthen the weak. That, Darwin complains, interferes with 'the survival of the fittest.' "[14]

Darwinism could have provided scientific justification for racism, but it rarely served that function. As John S. Haller Jr. has shown in his survey of evolution and racism in the late nineteenth century, Darwin scarcely affected racial discourse: "The views of the older pre-Darwinian concepts of racial inferiority remained essentially the same in the post-Darwinian period." In other words, the biblical account of God cursing Noah's son and grandson continued to provide ample sanction for believing in the inferiority of blacks. The historian Jeffrey P. Moran has recently concluded that "although leading evolutionists were often deeply involved in eugenics and scientific racism, African American anti-evolutionists never cited Darwinism's use by racists as an argument against the theory."[15]

Similarly, Darwinism, with its emphasis on sexual selection and its implication of female inferiority, might have been employed to justify gender discrimination, but it seldom was. As with business practices and racial attitudes, Christianity continued to provide more than ample justification for discrimination, beginning with Eve's subjugation to Adam. Kimberly A. Hamlin has recently shown that the few nineteenth-century feminists who invoked Darwin typically celebrated him for observing "that in all species, except among humans, females selected their sexual mates," thus justifying female choice. In addition, Darwinism invalidated the entire account of Adam and Eve.[16]

From time to time Darwinism was used to justify imperialism and war, but this, too, happened infrequently. Like the other ideologies we have examined, imperialism hardly needed scientific justification. Besides, even Herbert Spencer denounced imperialism as a "new barbarism."[17] Public perception of a linkage between Darwinism and war grew markedly during the Great War in Europe from 1914 to 1918, owing to allegations that Darwin's biology had played a role in persuading the Germans to declare war. Adopting the not-yet-familiar phrase "social Darwinism" for his title, David Starr Jordan (1851–1931), a prominent biologist and the founding president of Stanford University, dismissed the "dogma" that Darwinism justified war. "This 'biological argument for war,'" he declared, "has no scientific validity and no legitimate relation to the teachings of Darwin."[18]

Thus far I have focused on the limited ways in which Darwinism influenced social thought and behavior in the United States. I have not said much about social Darwinism, largely because few, if any, Americans mentioned it before the early twentieth century. Although scattered references to *Darwinisme social, Darwinismo sociale,* and *Sozialdarwinismus* began appearing in continental Europe about the early 1880s, the phrase "social Darwinism" seems to have first appeared in American publications in 1903, when the sociologist Edward A. Ross (1866–1951) applied it to those who saw "in the economic struggle a twin to the 'struggle for existence' that plays so fateful a part in the modification of species."[19] For decades thereafter, its meaning remained uncertain, though it typically carried a sinister connotation.

An early discussion at the 1906 meeting of the American Sociological Association revealed the confusion. The Dartmouth sociologist D. Collin Wells (1858–1911), perhaps the only American at the time to call himself a social Darwinist, presented a paper titled "Social Darwinism," explaining that he had borrowed the unfamiliar term from European writers to refer to "the gradual appearance of new forms through variation; the struggle of superabundant forms; the elimination of those poorly fitted, and the survival of those better fitted, to the given environment; and the maintenance of racial efficiency only by incessant struggle and ruthless elimination." The content of his talk surprised his commentator, Lester Frank Ward (1841–1913), the anti-Spencerian president of the association, who explained his confusion:

> In Europe, especially on the continent, there has been much discussion of what they call "social Darwinism." Not all scholars there agree as to what it is, but certainly none of them use the expression in the sense that Dr. Wells uses it. . . . Over there the discussion of this topic relates to two problems: first, the economic struggle, and, second, the race-struggle. Those who appear to defend this "social Darwinism" are biologists mainly and not sociologists at all. Most of the sociologists attack it.[20]

In the United States, in contrast, sociologists were virtually the only ones talking about social Darwinism.

The same year that Wells's article appeared, Ward damned his fellow sociologists for generally failing to understand "the true nature of the biological struggle" when writing about so-called social Darwinism. Many of them, he scolded, did not even recognize the distinction between Darwinian and Lamarckian evolution or understand the process of natural selection. But heedless of their ignorance and confusion, "certain of them have invented the phrase 'social Darwinism,' and have set it up as a sort of 'man of straw' in order to show their agility in knocking it down." In their eagerness to discredit suspect economic and racial theories, they had conjured up social Darwinism "in their own imagination" and set out to combat it "as valiantly as Don Quixote battled with the windmills."[21]

In 1944, the newly minted American historian Richard Hofstadter launched his career as a major interpreter of American culture with a slim volume titled *Social Darwinism in American Thought*. In it he attempted to show how the advocates of such conservative "social ideologies" as free trade, militarism, racism, imperialism, and eugenics had used Darwinism to advance their goal of rugged individualism.[22] Although he warned that it would "be easy to exaggerate the significance of Darwin for race theory or militarism either in the United States or in western Europe," many readers, including fellow historians, did exactly that.[23] With the publication of Hofstadter's monograph, "social Darwinism" became a trope in American histories, despite the accumulating evidence that its very existence was problematic.

It remains alive down to the present. The title of a recent essay by the historian of science Mark A. Largent says it all: "Social Darwinism Emerges and Is Used to Justify Imperialism, Racism, and Conservative Economic and Social Policies."[24]

III

TWENTIETH CENTURY

THAT THE MICHELSON-MORLEY EXPERIMENT PAVED THE WAY FOR THE SPECIAL THEORY OF RELATIVITY

Theodore Arabatzis and Kostas Gavroglu

Michelson and Morley found that the speed of the earth through space made no difference in the speed of light relative to them. The inference is clear . . . that all observers must find that their motion through space makes no difference in the speed of light relative to them. The above inference was clear, at least to Einstein, who . . . made it a cornerstone of his special theory of relativity.

—James Richards et al., *Modern University Physics* (1960)

What exactly do we mean by "myths in science"? Often we mean the propagation of stories that are at odds with the historical record—be it because their protagonists have specific views on how science has (or ought to have) developed or because teachers and textbook writers find them educationally expedient. When encountering such stories, historians of science have an obligation to dispel them by setting the record straight. In this chapter, however, we are interested in less blatant myths, the main characteristic of which is not the distortion of historical evidence. Rather, they slowly establish themselves by an appropriation of various aspects of the historical record, simplifying and transforming them for pedagogical, ideological, or philosophical aims. Within such a framework, we would like to examine the character of a rather idiosyncratic myth: the close—almost causal—connection that is stated in many textbooks, monographs, and popular writings between the null result of the Michelson-Morley (M-M) experiment and the emergence of the special theory of relativity (STR).

In the late nineteenth and early twentieth centuries, many well-known physicists were facing a rather awkward situation. A highly sensitive experiment, performed by Albert A. Michelson (1852–1931) and Edward W. Morley (1838–1923) in 1887 and designed to measure the relative velocity of the earth with respect to the ether, gave consistently null results. The ether was the necessary substratum of James Clerk Maxwell's (1831–1879) electrodynamics, since its prediction of the existence of electromagnetic waves required a medium through which they would travel. With the benefit of hindsight, the null result could have meant that the ether was not needed for the propagation of electromagnetic waves, but this was not something that could be entertained within the framework of nineteenth-century physics. All the explanations given for the null result were problematic or artificial: either the ether would have to possess contradictory properties or it would have unheard-of effects on the forces holding matter together, causing the contraction of material bodies along the direction of their motion.

The connection between the M-M experiment and the emergence of STR has been exposed as a myth, most notably by Gerald Holton (b. 1922), who made much of the fact that in Albert Einstein's (1879–1955) original 1905 paper on STR there is no explicit reference to the M-M experiment. Holton proceeded to a systematic examination of the origins of STR, concluding that the "role of the Michelson experiment in the genesis of Einstein's theory appears to have been so small and indirect that one may speculate that it would have made no difference to Einstein's work if the experiment had never been made at all."[1] Holton's assessment remained the same after new material surfaced in later years: "the influence of the famous experiment was neither direct and crucial nor completely absent, but small and indirect."[2] His influential and elegant debunking of that myth has stood the test of time and has been further refined by others who also examined additional archival material.[3]

Holton characterized the myth as a textbook account of the birth of relativity theory, heavily influenced by the positivist philosophy of science, which inflated the role of the M-M experiment as Einstein's main motivation. It is, indeed, the case that textbook writers have taken some liberties with Einstein's 1905 paper—and, as will be discussed further, they did the same thing with Max Planck's (1858–1947) 1900 paper on blackbody radiation as well as Niels Bohr's (1885–1962) 1913 papers on atomic structure. However, Einstein was not all that innocent in the development of the myth. Even though in his 1905 paper he remained silent about the M-M experiment, beginning in 1907, in a review article appearing that year, he was quite vocal and sometimes contradictory about the significance of that experiment.[4] Historians have been obliged to discuss this problem within the quagmire brought about by Einstein himself, not because of his silence about the role of this experiment in the formulation of the STR but because of his ambivalence about it for the rest of his life.

This qualifies as an idiosyncratic myth for two reasons: first, even though in the original paper of 1905 there is no mention of an explicit connection between the M-M experiment and the formulation of the STR, Einstein in subsequent papers, speeches, books, and interviews mentioned the M-M experiment and its positive role in the emergence and acceptance of the STR. Second, although historians of science do not altogether deny this connection, they do not consider the role of the null result to be as important as many writers of textbooks, monographs, and popular texts make it out to be, thus relegating the myth to something that is only partially false. Remarkably, there is virtual unanimity among Einstein scholars about the role of the M-M experiment in the creation of the STR: if it did play a role, it was by no means significant.[5] Our task, then, is less to set the record straight than to examine the character and implications of the M-M myth.

The history of the relation between the M-M experiment and the STR has an aspect not sufficiently emphasized by Holton.[6] In 1907, Einstein, in the aforementioned review article on "The Relativity Principle and the Conclusions Drawn from It," made repeated references to the M-M experiment and emphasized that it suggested "the simplest possible assumption," that is, the principle of relativity.[7] Holton mentions this review article only in one footnote and refers to Einstein's allusions to the M-M experiment with the cryptic phrase that "here again we find a sequence of sentences which can be considered implicit history."[8] It is also worth noting that in 1911, Max von Laue (1879–1960)—who had visited Bern to meet Einstein while he was still in the patent office and who wrote the first monograph on relativity—claimed that the M-M experiment was "the fundamental test for the theory of relativity," thus giving a green light to subsequent textbook writers for underlining the importance of the M-M experiment in the formulation of the STR.[9] In view of these statements, which may have set the tone for subsequent textbook writing, the "textbook myth" of the origins of relativity starts to look justifiable, especially within the framework of the pedagogical aims of textbook writers.

It appears, then, that the authors of scientific texts and popular writings "accused" of creating and propagating a myth have very little to be apologetic about. They could have a perfectly legitimate answer: we have just been following what the master himself said and wrote on various occasions, such as in his 1907 review article and in his excursions into the history of physics.[10] In fact, the situation is even more complicated. Einstein was, throughout his life, of two minds regarding the role of the M-M experiment in the origins of STR. He appeared undecided as to what the true story was, at times denying that he knew about it and at times giving it a decisive role; thus, he created an ideal framework for the perpetuation of the myth.

Might it be the case, then, that it is the historians of science who are off the mark in debunking a nonexistent myth? Might it be the case that they are, at least partly, responsible for creating another myth in response to the myth they attribute to textbook writers? The answer is yes and no. To the extent that textbook authors and popularizers have overblown the significance of the M-M experiment for the genesis of the STR, the answer is yes, these authors have created a myth. To the extent that they have tried to reconstruct, in a pedagogically plausible way, the origins of STR, focusing on most of what Einstein wrote about it *after* 1905, the answer is no.

This is basically the situation regarding the myth involving the M-M experiment and the origins of the STR. However, most textbook authors have contributed decisively to the creation of another myth, which has shaped rather strongly the historical consciousness of many generations of physicists: in many textbooks, the chapter on relativity is part of the treatment of mechanics and rarely covers topics from electromagnetism. The STR is presented as the "correct" mechanics, which reduces to Newtonian mechanics when the speed of light approaches infinity. This is what physics students have been taught, and this is the "feeling" of the overwhelming majority of physicists. This, then, is a "real" myth, having very little to do with Einstein and the history of the STR. Einstein formulated the STR in response to problems in electrodynamics, and one of his main aims was to get rid of "asymmetries" in electromagnetic theory. Furthermore, when he chose to speak of the M-M experiment, Einstein emphasized, time and again, its significance for supporting the principle of relativity—*not* the constancy of the speed of light. But in many textbooks one finds a different inference: that the M-M experiment demonstrated the constancy of the speed of light regardless of the motion of its source.

Take, for instance, two textbooks that have helped educate hundreds of thousands of university students and have provided a basic knowledge and the overall framework of modern physics.

They have been repeatedly revised, reprinted, translated, and used around the world. In both of them, the STR is introduced toward the end of the section on mechanics, with no reference to electrodynamics, and the null result of the M-M experiment is related to Einstein's postulate about the speed of light:

> The null result of the Michelson-Morley experiment to detect the drift of the earth through an ether . . . can only be understood by making a revolutionary change in our thinking; the new principle we need is simple and clear: The speed of light is independent of the motion of the light source or receiver.[11]

> The result of the Michelson Morley experiment . . . shows that the velocity of light is the same in all directions on the moving earth.[12]

Here, then, are the ingredients of another myth that has been systematically cultivated by textbook writers: (1) that the M-M experiment proved the constancy of the velocity of light, regardless of the velocity of its source; (2) that the M-M experiment refuted the ether hypothesis; (3) that special relativity was developed in the context of mechanics rather than electrodynamics; and (4) that special relativity was a generalization of Newtonian mechanics.

In such physics textbooks, the M-M experiment is presented in the context of "pedagogical reconstructions" of relativity theory, the main aim of which is to facilitate the comprehension of the theory and to present a plausible and straightforward account of its origins (or its evidential base). Of course, these reconstructions are not historiographically innocent. To serve their purpose, they inevitably oversimplify and often distort the historical record. Furthermore, these reconstructions are framed in terms of an intuitively plausible "schema" of how new theories are born and come to be accepted, which consists in "putting the phenomena first": new theories are motivated by puzzling phenomena that cannot be explained by those theories' predecessors.[13] Besides accounting for puzzling phenomena, new theories always include their predecessors as limiting cases, which hold within a well-circumscribed domain. Thus, the continuity and progress of scientific change is secured.

Such a schema often captures the way all of modern physics is presented in textbooks. It expresses the collective aim of textbook writers to articulate a "canonical" textbook historiography. Consider, for instance, Planck's 1900 paper, in which he is preoccupied with the second law of thermodynamics and entropy. He discusses the experimental measurements for blackbody radiation and the difficulties in providing a satisfactory theoretical explanation for a wide range of the emitted frequencies; he then proceeds to derive Wien's law, announced in 1893 by the German physicist Wilhelm Wien (1864–1928), who proposed it in an empirical manner. Nowhere in the paper is there any mention of the Rayleigh-Jeans law for low frequencies. However, textbook accounts of how Planck introduced his theory suggest a different approach: They first present blackbody radiation and discuss Wien's law and its success in accounting for high frequencies. They then discuss the Rayleigh-Jeans formula for low frequencies. And they finally present Planck's work as providing a unified account of the whole spectrum of blackbody radiation. Besides the fact that the Rayleigh-Jeans law is not mentioned in Planck's paper, James Jeans's (1877–1946) improved version of Lord Rayleigh's (1842–1919) derivation did not appear until 1905.[14]

Similar problems beset textbook accounts of Bohr's model of the atom. Bohr, in the first part of his 1913 trilogy "On the Constitution of Atoms and Molecules," starts with the paradox that within the framework of classical electrodynamics, Ernest Rutherford's (1871–1937) model of the atom cannot account for the stability of matter. This is what Bohr tries to alleviate, and only at the end of his paper does he show that his model, which guarantees the stability of atoms, also provides a very satisfactory account of the hydrogen spectrum. Textbooks, however, provide a different picture. Almost invariably they start with the empirical

Paschen-Balmer formulas for the hydrogen spectrum, and present Bohr as aiming directly at their explanation.

Ideas explicitly expressed in Einstein's, Planck's, and Bohr's original papers are "tampered with" in order to fit into the textbook culture of pedagogical expediency. The works of Einstein, Planck, and Bohr are presented as being motivated by their wish to explain experimental results that had been either unaccounted for or explained in an unsatisfactory manner. In textbook accounts, Einstein appears dissatisfied with the Lorentz-Fitzgerald explanation of the M-M experiment; Planck not satisfied with the lack of explanation for the Wien and the Rayleigh-Jeans laws; and Bohr not content with the hitherto unexplained Paschen-Balmer series. It appears that textbooks have followed a coherent historiographical viewpoint, which, while differing in its details, has an incredibly strong grip on both writers and readers: the heroes of modern physics are those who provided ingenious explanations of recalcitrant experimental results.

THAT THE MILLIKAN OIL-DROP EXPERIMENT
WAS SIMPLE AND STRAIGHTFORWARD

Mansoor Niaz

By a brilliant method of investigation and by extraordinarily exact experimental technique Millikan reached his goal.

> —Allvar Gullstrand, Presentation speech to award Nobel Prize to Robert A. Millikan (1924)

[The] Millikan oil-drop experiment [was the] first direct and compelling measurement of the electric charge of a single electron. It was performed originally in 1909 by the American physicist Robert A. Millikan, who devised a straightforward method of measuring the minute electric charge that is present on many of the droplets in an oil mist. . . . Through repeated application of this method, the values of the electric charge on individual oil drops are always whole-number multiples of a lowest value—that value being the elementary electric charge itself (about 1.602×10^{-19} coulomb). From the time of Millikan's original experiment, this method offered convincing proof that electric charge exists in basic natural units.

> —"Millikan's Oil-Drop Experiment," *Encyclopedia Britannica*

What is electricity? This seems to be a simple question and may be discussed these days even in a primary-school classroom. The manifestation of electricity can be demonstrated by the movement of a pith ball when brought in contact with a glass rod rubbed with silk. Around 1750, Benjamin Franklin (1706–1790) was perhaps the first to suggest the concept of electrical particles or atoms. In 1881, George Johnstone Stoney (1826–1911) first made an estimate of the ultimate electrical unit and named it the *electron*. Joseph John Thomson (1856–1940) presented experimental evidence based on cathode-ray experiments that led to the determination of the charge to mass ratio of particles, which were later recognized as universal charged particles and finally led to the discovery of the electron.

The determination of the elementary electrical charge *(e)* aroused considerable interest in the scientific community and prompted Robert A. Millikan (1868–1953) to become deeply involved in its measurement. Millikan's oil-drop experiment is generally considered to be a simple, beautiful, and straightforward experiment that unambiguously led to the determination of the elementary electrical charge. According to a poll conducted for *Physics World,* its readers considered the oil-drop experiment to be one of the ten "most beautiful" experiments of all time. In this chapter, I explain that the experiment was not only difficult to perform but also elusive with respect to its interpretation, leading to considerable controversy in the scientific community, which lasted for many years.[1]

From Pith Balls to Water Droplets to Oil Droplets

Physical scientists consider the elementary electrical charge *(e)* to be an important milestone in understanding the electrical nature of matter. However, some of the early experiments, such as those done by Thomson at the Cavendish Laboratory in Cambridge, were extremely difficult to design. Early researchers often conducted their experiments in cloud chambers in which they could observe clouds of charged water droplets moving in electrical and gravitational fields. These experiments were subject to various errors, such as the evaporation of the water droplets and the instability, distortion, and lack of sharpness of the surface of the cloud, which made measurements difficult and uncertain. In 1906, when

Millikan began performing similar experiments at the University of Chicago, he faced the same difficulties. Thus, he innovated by using a 10,000 V battery instead of the 4,000 V battery used in previous experiments. This led to an entirely unexpected result: clouds of water droplets disappeared while a small number of individual drops remained, which could be easily observed as bright points by illumination with light. This brought the decade-long technique of measuring electrical charges by the formation of clouds to an abrupt end.[2]

Using his new technique, Millikan obtained results that he presented at the meeting of the British Association for the Advancement of Science held in Winnipeg (Canada) in August 1909. Despite his partial success, several sources of error—such as the gradual evaporation of the individual water droplets, the lack of uniformity of the electric field, and the difficulty of holding a drop under observation for more than a minute—remained. Thus, Millikan introduced various changes in his experiments, the most important of which was arguably the replacement of water with oil (thus avoiding rapid evaporation). Millikan later recalled that the idea of using oil instead of water occurred to him suddenly while he was riding the train back to Chicago from the Winnipeg meeting, where he had met Ernest Rutherford (1871–1937), a pioneer in nuclear physics, and others working on similar problems. With the use of oil, Millikan achieved a series of discoveries: (1) oil drops act essentially like solid spheres; (2) the density of the oil drops is the same as that of the oil in bulk; and (3) oil produces a better estimate of the frictional force exerted on spherical drops as they move in the surrounding fluid. He published these results in 1913.[3]

Controversies

Felix Ehrenhaft (1879–1952), working at the University of Vienna, conducted experiments based on metal drops (rather than oil), which were quite similar to those of Millikan. In fact, both scientists obtained experimental data that were quite similar. Millikan, however, postulated the existence of a universal charged particle (the electron), whereas Ehrenhaft postulated the existence of subelectrons based on fractional charges. The controversy started when Ehrenhaft recalculated Millikan's data from oil drops and found a large spread of values of the electrical charge, quite similar to his own data. Ehrenhaft showed how Millikan's method led to paradoxical situations. For example, Millikan considered two oil drops having very similar charges to have different numbers of electrons. Colleagues wondered how to explain these differences. The controversy between the two lasted for many years, from around 1910 to 1923, when Millikan was awarded the Nobel Prize. Recently opened archives for the Nobel Prize show that although Millikan was a nominee for the prize from 1916 on, the prize committee recommended that it be withheld as long as the controversy with Ehrenhaft continued.[4]

Almost fifty-five years later, in 1978, Gerald Holton (b. 1922) added a new dimension to the controversy with his discovery of Millikan's two laboratory notebooks at the California Institute of Technology. The Millikan notebooks have 175 pages of data from experiments conducted between October 28, 1911, and April 16, 1912, much of the content used in his *Physical Review* article in 1913. In these notebooks, Holton found data from 140 drops, but the published article reported results from only 58 drops. What happened to the other 82 drops? It seems that Millikan made a rough calculation for the value of e as soon as the data for the times of descent/ascent of the oil drops started coming in and ignored any experiment that did not give the value of e that he expected. More recently, after having read a preliminary version of this essay, Holton added, "So even if Millikan had included *all* drops and yet had come out with the same result, the error bar of Millikan's final result would not have been remarkably small, but large—the very thing Millikan did not like."[5]

This leads to the question: What was the warrant under which Millikan discarded more than half of his observations? Millikan's guiding assumption, based on the atomic nature of electricity and the value suggested by the previous experiments of Ernest Rutherford at the University of Manchester, was a constant source of guidance. From his own experiments, Millikan had learned that all data could not be used because of difficulties associated with evaporation, sphericity, radius, change in the density of drops, and variation in experimental conditions (battery voltages, stopwatch errors, temperature, pressure, and convection). Like Millikan, Ehrenhaft had also obtained data that he interpreted as integral multiples of the elementary electrical charge *(e)*—as well as data for many drops that did not lead to an integral multiple of *e*. According to Holton, Ehrenhaft used data from all the drops that he studied, producing the impasse with Millikan. "It appeared," Holton concluded, "that the same observational record could be used to demonstrate the *plausibility of two diametrically opposite theories,* held with great conviction by two well-equipped proponents and their collaborators."[6] Furthermore, Holton showed that (contrary to many commentators and textbook authors) Millikan had not measured the charge on the electron itself but rather the transfer of charge on drops as an integral multiple of the elementary electrical charge *(e)*.[7]

The Oil-Drop Experiment in Textbooks and Laboratories

The oil-drop experiment is an important part of high school and introductory university physics and chemistry courses in almost all parts of the world. Thus, the Millikan-Ehrenhaft controversy can open a new window for students by demonstrating how two well-trained scientists could interpret the same data in two different ways. Studies based on thirty-one general chemistry textbooks and forty-three general physics textbooks published in the United States showed that none of the textbooks referred to the controversy; very few of them explained that Millikan did not measure the charge on the electron but rather the transfer of charge; and very few of them explained satisfactorily that the oil-drop experiment was extremely difficult to perform because of the incidence of various experimental variables.[8]

Most textbooks ignored one of the most important aspects of the oil-drop experiment—namely, the guiding assumptions of both Millikan and Ehrenhaft. Relying on previous research and historical antecedents, Millikan embraced the atomic nature of electricity. In contrast, Ehrenhaft adopted the anti-atomistic ideas of Ernst Mach (1838–1916) and hence advocated the existence of fractional charges (subelectrons). The textbook accounts similarly ignored various other essential elements of the oil-drop experiment, such as how scientists persevere in the face of difficulties, how they find new alternative interpretations, and how they deal with criticisms from their peers that lead to controversies. On the contrary, these textbooks endorse a vision of science that leads to the perpetuation of such myths as those about an accurate direct measurement of the charge on the electron, about developing a series of brilliant experiments which measured the elementary electrical charge of an electron, and about measuring the charge on the electron directly and forming the basis of measurements of great precision of this quantity. Interestingly, textbooks have continued to ignore the controversial nature of Millikan's data-reduction procedures even after the publication of Holton's study in 1978.[9]

In many parts of the world, the oil-drop experiment still forms an important part of laboratory instruction in undergraduate physics. One study based on eleven laboratory manuals published in the United States reported that they closely followed the portrayal of the experiment that appears in general physics and chemistry textbooks. However, some instructors and authors of manuals did refer to the difficulties involved in replicating the experiment—for example, in choosing which drops to observe. Probably many teachers would be surprised that Millikan himself faced the same dilemma, in one case discarding almost 59 percent of the drops studied.[10]

It seems that the oil-drop experiment continues to be considered in present-day textbooks as a simple, beautiful, precise, brilliant, and clever one, which provided convincing evidence for the determination of the elementary electrical charge *(e)*. Interestingly, present-day undergraduate students do not consider the experiment either simple or beautiful but find it rather frustrating.

THAT NEO-DARWINISM DEFINES EVOLUTION
AS RANDOM MUTATION PLUS NATURAL
SELECTION

David J. Depew

Evolution in the sense of common ancestry might be true,
but evolution in the neo-Darwinian sense—an unguided,
unplanned process of random variation and natural selection—
is not.

>   —Cardinal Christoph Schönborn, "Finding Design in
>   Nature" (2005)

Since the 1940s, the principles of what has been called neo-Darwinism (or the modern evolutionary synthesis) have guided professional evolutionary inquiry.[1] These principles stemmed from the fusion of Mendelian genetics with Darwin's idea of natural selection. Therefore, it might seem natural enough to summarize neo-Darwinian evolution as random genetic variation plus natural selection. This is, in fact, how antievolutionists prefer to characterize it. For instance, the Discovery Institute, an organization that promotes the intelligent-design version of antievolutionism, says it is "skeptical of . . . neo-Darwinism's . . . claims for the ability of random variation and natural selection to account for the complexity of life."[2]

This way of construing neo-Darwinism might appear to have spread to the Roman Catholic Church.[3] Christoph Schönborn (b. 1945), the cardinal archbishop of Vienna and a former student of retired Pope Benedict XVI (b. 1927), wrote the epigraph to this chapter in an op-ed piece in the *New York Times* in 2005. Some Catholic evolutionary biologists soon intervened, however, to deflect the church from supporting this way of characterizing neo-Darwinism. More factors than mutation and selection, they pointed out, affect evolutionary change, and therefore the portrayal of evolution as a process in which accidents are selectively preserved or eliminated is a myth.[4] Chance variation certainly plays an essential role in the evolution of adaptations, which Darwinians as much as their creationist opponents recognize as a prominent feature of organisms. However, adapted traits and properties, which are so functional and goal directed that they seem intentionally designed, are also the outcome of so many other factors that "random genetic variation plus natural selection" is an inadequate description of how adaptive evolution works.

A good way to clear up misunderstandings about neo-Darwinism is to relive some of its history. Prior to the end of the nineteenth century, little was known about the mechanisms of heredity. However, in the 1880s, the embryologist August Weismann (1834–1914) began showing that only germ-line factors are heritable and that bodily characteristics acquired in the course of a lifetime cannot be passed on to descendants. Weismann's account of inheritance gave rise to neo-Darwinism in the original and most proper sense of the term. Neo-Darwinians were "neo" because they explained adaptation solely in terms of natural selection's elimination of unfit and preservation of fit variants in germ-line factors—unlike Darwin himself, who acknowledged use and disuse as an auxiliary explanation of some adaptations and assumed that acquired bodily characteristics could be passed on to descendants (see Myth 10). Weismann proposed "blastogenesis" in the place of Darwin's "pangenesis" in order to insist that only characters in the germ line and not in the whole body are heritable.[5]

One might imagine that the (re)discovery in 1900 of what Mendelians thought they saw in Mendel's neglected work on hybridization (see Myth 16) would have supported neo-Darwinism. At the time, however, this was far from the case. Weismann's germ-line factors, it now appeared, came in discrete unblended but combinable units. Unless they spontaneously mutated, these heritable factors would remain unchanged in gene pools generation after generation, especially when they were recessive rather than dominant and were thus shielded from natural selection. Early neo-Darwinians held that in addition to eliminating unfit germ-line variants, natural selection could slowly spread mutations that kept lineages adapted to their environments. Unfortunately, they also assumed that the adaptive effects of a markedly superior variant in one generation would progressively wear off in later generations, leaving descendants with characteristics that were no better than average and therefore with no net gain in adaptedness.

Early Mendelians—such as William Bateson (1861–1926), Hugo de Vries (1848–1935), and Wilhelm Johannsen (1857–1927), who coined the term "gene" in 1909[6]—spotted the tensions in this early neo-Darwinian view. They were skeptical of the power of natural selection ever to beat the tendency of traits to regress to an undistinguished mean. Instead, they relied on sudden, single-leap mutations that just happened from the start to be adaptive. Accordingly, for these Mendelians, mutation—not natural selection—was evolution's "creative factor."[7] In response, neo-Darwinians pointed out that single-shot mutations were overwhelmingly likely to disrupt the adaptedness on which the viability of organisms depends and so would be summarily purged by natural selection. Therefore, at the beginning of the twentieth century, it would have appeared not just dubious but incoherent to say that evolution consists of random mutation plus natural selection.[8]

Eventually, findings in genetics did support the unification of mutation and selection, but it took a few decades for this fusion to be accomplished.[9] The breakthrough came in 1918, when the statistical genius Ronald A. Fisher (1890–1962) used a mathematical theorem first derived in 1908 to demonstrate that when Mendel's laws are expanded to whole populations of freely interbreeding organisms, the supposed tendency of new variations to regress to an undistinguished mean is nowhere to be found.[10] From this perspective, Mendelians were right to hold that unexpressed "alleles" (from Greek for "other" or "alternative") remain in gene pools as recessives. But neo-Darwinians were right to doubt whether single-shot mutations with large effects are evolution's innovative factor. Mutations of this sort are statistically highly improbable, and so it is even less probable that a concerted sequence of lucky mutations will drive a lineage toward greater adaptedness. In fact, it would be almost as miraculous as advocates of intelligent design hope it is. Even though unification between neo-Darwinian adaptationists and mutationists occurred a century ago, the implications of their integration are still widely misunderstood.

What, then, is the creative factor in evolution? When empirical examination of how organisms actually live is combined with statistical-probabilistic thinking about populations, we can see that mutations, each with a small effect, and natural selection can indeed slowly move populations away from an equilibrium distribution of alleles that in the absence of these factors would remain the same generation after generation.[11] So natural selection can be as innovative as Darwin thought, though usually in combination with other factors. In addition to mutation, there can be gene flow, when migrants spread variants by moving from one subpopulation to another, interbreeding as they go. Genetic drift, too, can occur, in which new variants can get a toehold purely by chance in the small populations in which many species live. Drift occurs in small populations for the same probabilistic reason that in roulette a ball may land six or seven times on red without violating the expectation that in the long run the number of red and black landings will even out.[12] If these variant alleles are adaptively favorable, natural selection can then spread them to and through an interbreeding population.

From the statistical-population perspective, we can also see that the genetic variation that natural selection uses to fuel the process of adaptation is not exclusively mutational. Although mutation in the genetic material is a source of variation, the recombination of genetic material during meiosis—the cell division during which sperm and egg are produced—provides so much of the proximate variation on which natural selection works that, as Theodosius Dobzhansky (1900–1975), one of the founders of the modern evolutionary synthesis, remarked, "Suppression of the mutation process . . . would probably have little effect on the evolutionary plasticity of a population for some time to come."[13] In recent decades, even more sources of variation have come into view, as the genetics of the developmental process has become better understood. Gene sequences that regulate the timing and rate of developmental processes are a source of heritable variation, making it clear that natural selection affects traits because more fundamentally it affects developmental trajectories. This discovery shows how large morphological changes can result from minor genomic changes.[14] There are also sources of epigenetic heritable variation, such as chemical side chains that attach themselves to DNA (methylation) and more.[15]

Population thinking, accordingly, enables us to appreciate that natural selection does more than just eliminate unfit accidental variation. It selects *against* harmful inherited factors, to be sure, but working on variation of several kinds, it also selects *for* traits of organisms, populations, groups, and species that *become* adapted as natural selection, working in combination with other factors, amplifies reproductively more successful variants through interbreeding populations. In this process, natural selection acts in several modes. It acts disruptively, so that populations of the same species exploit a slightly different resource base and eventually become reproductively isolated, evolving into new species. In stable environments it acts directionally, favoring one among several genotypes if the former happens to confer an advantage to its bearers. In unstable environments it "balances" currently adaptive genotypes with others that may be useful when circumstances change by preserving the respective variants as recessive alleles. In sum, as neo-Darwinism matured it showed natural selection to be the preeminently creative factor in evolution that Darwin thought it to be.[16]

Based on all of the above, it becomes clear that attempts to describe neo-Darwinism as "random mutation plus natural selection" fail to do justice to a framework of inquiry that has guided the discovery of scientific knowledge for well over half a century. It is possible, however, that some still prefer this formula because this way of describing neo-Darwinism's limited but real reliance on chance makes it look less persuasive than intelligent design. "How can all these wonderful adaptations be nothing more than preserved accidents?" ask supporters of intelligent design. They can't, Darwinians reply—nor should the work of natural selection and related factors be looked at that way.

The confusion stems from different understandings of what is meant by the term "random." "Random" does not mean "haphazard" or "open to all possibilities" but rather "unintentional" and "unpredictable." When Darwin spoke of variation as "chance," he meant only that the causes of variation are unrelated to its subsequent adaptive utility, not that variation has no causes. In fact, Darwin and his early supporters presumed that variation is caused by unknown deterministic laws of physics and chemistry, like those at the heart of the science of their day. It was only after molecular geneticists discovered that spontaneous changes in DNA sequences are a primary source of mutations that scientists could entertain the notion that mutation is random in the strong sense of being stochastic.

This created an ambiguity. Present-day Darwinians still use the term "chance" the way Darwin did. The genetic variation that is relevant to natural selection arises independently of any effect it might have on reproductive success. In this sense, "chance variation" is part of the definition of natural selection. However, in positioning "random genetic variation" as prior to and independent of the process of natural selection, the formula "random genetic variation plus natural selection" slyly leads students and others to suspect that, as one anti-Darwinian put it, "According to Darwinism our existence is a mere accident."[17] That is because the formula stresses natural selection's weeding-out role; renders invisible its "creative" role in adapting organisms to environments; greatly shortens the long chain of causes and levels between mutations and traits; and overlooks Darwin's appeal to natural processes with unpredictable outcomes to explain, not explain away, the functional, goal-directed, and purposive characteristics of organisms.

If the adapted traits of adapted and coadapted organisms were products of conscious design, we would expect a good designer, like a good engineer, to have a larger end in view, such as the evolution of humans. Darwinians debate how much of the history of life on earth is adapted and how much is accidental, but they can't satisfy demands for a history that is guided toward the appearance of our species.[18] This can be disheartening to those who ask biology to slake their thirst for overall evolutionary purpose. In their disappointment, they may underestimate the vast array of functional parts and goal-directed behaviors that neo-Darwinians *can* vouch for. Darwinians can help correct this misperception by declining to characterize adaptive natural selection as "design without a designer," which itches for a fight it need not have.

MYTH 21

THAT MELANISM IN PEPPERED MOTHS IS NOT A GENUINE EXAMPLE OF EVOLUTION BY NATURAL SELECTION

David W. Rudge

> So is camouflage the reason natural selection favored the dark moths? Probably not. Further work showed these moths don't spend much of their days on tree trunks. Some other effect of pollution seems to be at work.
>
> —George B. Johnson and Jonathan B. Losos, *The Living World* (2010)

The phenomenon of industrial melanism refers to a rapid increase in the frequency of dark forms of moths in the vicinity of industrial areas, which occurred in the wake of the Industrial Revolution (about 1760–1840). It was first noticed in the peppered moth, *Biston betularia,* a common moth known throughout Britain and Continental Europe and named for its pale, speckled appearance. The discovery of a rare dark form of the moth in 1840 near Manchester, England, led naturalists and insect collectors to search for more examples of what was initially regarded as a sport of nature. Over a period of just a few decades, they discovered numerous additional examples of dark peppered moths and also dark forms of many other species. What made it particularly curious was that the rise in the frequency of the dark form in these species was confined to areas in the vicinity of manufacturing centers, where large-scale air pollution had visibly darkened the environment. Geneticists determined that the dark form of the peppered moth was the result of a variation in a single gene, and in an influential paper, John Burdon Sanderson "Jack" Haldane (1892–1964) pointed out that the speed of the apparent spread of this gene implied that it had an enormous selective advantage in the affected environments.[1]

These considerations led evolutionary biologists to recognize that the phenomenon of industrial melanism was an example of natural selection taking place before their eyes. This becomes obvious when we consider Charles Darwin's (1809–1882) theory of natural selection with reference to the phenomenon of industrial melanism:

1. If the peppered moth has both dark and light forms, and if these differences are correlated with survival differences in different environments; and,
2. If the dark and light forms are heritable; and,
3. If there is a competition in nature for resources, owing to the fact that the moths reproduce far in excess of those that can possibly survive; then,
4. It follows that the form of the moth that is correlated with an increased chance of surviving in an environment will increase in frequency in the population inhabiting that environment over time (if it is not already in equilibrium).

The reader should notice that Darwin's theory as stated here does not require us to know precisely how the gene responsible for dark coloration affords dark moths a greater survival advantage in polluted environments. The mere fact that dark coloration is inherited, and that this difference is somehow correlated with survival differences, makes this an example of natural selection.

The claim that industrial melanism is an example of natural selection is not in dispute among scientists who research the phenomenon, although precisely why the dark form became more common in affected areas has historically been the subject of a great deal of debate. James William Tutt (1858–1911) is often identified as the first to popularize the idea that the dark form was becoming more common in the affected areas because of the camouflage value of being dark when resting on soot-darkened backgrounds against visual predators, such as birds. And, indeed, when one compares the pale and dark forms of the peppered moth resting on pale lichen-covered bark taken from an unpolluted forest and the two forms resting on soot-darkened bark, this intuition seems obvious.

Early investigators also considered other explanations for why the dark form was becoming more common. Edmund Brisco "Henry" Ford (1901–1988), one of the pioneers of ecological genetics (an experimental branch of evolutionary biology devoted to the study of the genetic basis of evolution by means of laboratory and field studies), was convinced that the phenomenon was more complicated than most biologists assumed. Ford emphasized anecdotal evidence showing that the dark form in affected species invariably appeared to be "hardier." Precisely what this meant appears to have depended on the investigator, but the consensus was that the dark form was physiologically superior to the pale form—that is, it was better able to tolerate toxins present in the pollutants. Ford concluded that this physiological advantage was the primary reason why the dark form was becoming more common in the affected areas, and that the spread was limited to polluted areas owing to the obvious handicap of dark coloration in unpolluted areas, where the dark form would be quite conspicuous when it rested against a pale lichen-covered background. James William Heslop Harrison (1881–1967), another British naturalist in the early twentieth century, advanced a different explanation. On the basis of experimental studies in which he fed caterpillars with contaminated foliage, he claimed that lead salts present in the soot fallout from industry had mutagenic properties. These investigations caused quite a stir at the time they were published because they represented evidence of so-called Lamarckian inheritance (the possibility that traits acquired by parents can be passed on to their offspring; see Myth 10). Other investigators attempted but failed to duplicate Harrison's results, which led to allegations that his original investigations were fraudulent.[2]

Part of the reason why early investigators were initially reluctant to seize on Tutt's altogether intuitive explanation was that it was not clear that birds were significant predators on the moths, nor that they would have the same difficulty humans have when attempting to spot them against their matching backgrounds. In the early 1950s, Henry Bernard Davis Kettlewell (1907–1979) conducted a series of pioneering field experiments aimed at resolving these issues. His experiments relied on a technique of mark, release, and recapture, whereby he released known quantities of marked pale and dark moths in both polluted and unpolluted areas of the countryside and then attempted to recapture as many as possible by means of light and assembling traps. He reasoned that all things being equal, the recapture rates for the two forms should be quite similar. If, on the contrary, one form was at an advantage in a particular environment (for example, the dark form was better able to escape from bird predators than the pale in a soot-darkened environment), the recapture rate for the favored form would be higher, because more had survived during the interval between release and recapture. And this is precisely what Kettlewell found: in the polluted setting, the recapture figure for the dark form was twice that of the pale form; and conversely, in the unpolluted setting, the reverse was true. Kettlewell also placed moths representing both forms on soot-darkened and pale lichen–covered tree trunks in the two settings and had an associate, Nikolaas "Niko" Tinbergen (1907–1988), film the order of predation from behind a hide (a barrier that hid his presence from the birds). These films documented that a variety of birds do prey on the moths and that they do so with reference to how conspicuous the moth is when it rests on the tree trunk. They also documented the speed with which birds capture moths, explaining why bird predation had been previously unnoticed.[3]

The intuitive nature of the phenomenon (compared to other examples available at the time) and the elegant simplicity of Kettlewell's apparently definitive demonstration led to its wholesale adoption by textbooks. By the late 1960s, industrial melanism had already become ubiquitous in American biology textbooks, leading to its designation as *the* classic demonstration of natural selection.[4] This reaction contrasts with that of other researchers on the phenomenon, who while embracing Kettlewell's general conclusion nevertheless expressed reservations about the conduct of his investigations, such as how Kettlewell released moths onto tree trunks, and his assumption that moths rest on tree trunks in plain sight during the day. Indeed, one can say that much of the research on industrial melanism since Kettlewell's first investigations in the early 1950s, by him and others, has been an attempt to remedy these perceived problems. It should be recognized, however, that the basic outline of the explanation we associate with Kettlewell has been confirmed by at least eight field studies. An impressive study conducted on two continents documented a similar rise and predictable fall in the frequency of dark peppered moths in Britain and the United States following the advent of clean-air legislation. Moreover, a recent, large-scale six-year predation experiment specifically designed by the late Michael Majerus (1954–2009) to address perceived problems associated with the conduct of Kettlewell's original investigations provides the most direct evidence yet for the selective role of bird predation. There is simply no doubt among researchers who work on the phenomenon that it is a dramatic example of natural selection that has occurred primarily as a result of differential bird predation. Contemporary research has established that the phenomenon is more complicated than textbooks imply by drawing attention to the role of other factors, such as sulfur dioxide concentrations and differential migration.[5]

Critics of the standard story we associate with Kettlewell often suggest that recent anecdotal observations of moths resting higher in the canopy, rather in plain site on the surface of tree trunks, completely undermine this example. Scientists who work on the phenomenon draw no such conclusion. This is not because they are being dogmatic in the face of contrary evidence but because an explanation primarily in terms of selective bird predation remains the best account given the available evidence. Chance observations of moths resting elsewhere simply draw our attention to the fact that more systematic research on the life history of the moth is needed. That some outstanding questions remain regarding why the gene responsible for dark coloration confers a selective advantage does not call into question that this is—and remains—a particularly well-documented example of natural selection.

The phenomenon of industrial melanism has nevertheless in recent years become a lightning rod for critics of evolution, who draw attention to differences between textbook accounts written for introductory audiences and the subtleties shared among scientists who write for technical publications. The intelligent-design theorist Jonathan Wells (b. 1942) harps on these discrepancies as indicative of the systematic way that textbook writers misrepresent the evidence for evolution, urging that biology textbooks be equipped with warning labels.[6] He is particularly exercised by the use of "staged" photographs, which misleadingly suggest to the reader that the moths are known to rest on tree trunks (ignoring the more obvious explanation that the textbook is simply trying to illustrate how difficult or easy it is to spot the moth when it rests on a matching or contrasting background). Judith Hooper (b. 1949) in a recent popularization all but accuses Kettlewell of committing fraud, a completely baseless accusation.[7]

As a predictable consequence of these unfounded attacks, there has been a dramatic decline in the use of industrial melanism as an example of natural selection in American biology textbooks since the early 2000s.[8] Part of the problem reflects a general limitation of all textbooks. Textbook writers, in consideration of space limitations and their intended audience, present science as briefly and simply as possible. This systematic omission of details regarding the process of science has the unfortunate consequence of portraying the results of science as certain, rather than tentative and the object of continued investigation. It also perpetuates a myth that Kettlewell worked in isolation, but in truth he relied heavily on other colleagues, not to mention a veritable army of amateur naturalists who collected records of the distribution of pale and dark forms.[9] It is important to recognize that every fact in science has a story behind it, and indeed it is by comparing the certitude with which a textbook entry presents industrial melanism with what is actually known about the phenomenon that one can begin to appreciate aspects often referred to as the nature of science, such as the tentative nature of scientific knowledge.[10]

THAT LINUS PAULING'S DISCOVERY OF THE

MOLECULAR BASIS OF SICKLE-CELL ANEMIA

REVOLUTIONIZED MEDICAL PRACTICE

Bruno J. Strasser

It has taken us almost half a century to cross the threshold from laboratory to clinic with respect to this genetic anemia, the "first molecular disease."

> —A. N. Schechter and G. P. Rogers, "Sickle Cell Anemia: Basic Research Reaches the Clinic" (1995)

How Are Human Disorders Caused by Single Genes Inherited? . . . Sickle-cell anemia, a recessive disease in which defective hemoglobin is produced, results from a specific mutation in the hemoglobin gene.

> —Teresa Audesirk, Gerald Audesirk, and Bruce E. Byers, *Biology: Life on Earth* (2011)

There is a gene for breast cancer and one for Alzheimer disease and yet another one for obesity and one for alcoholism. The view that single genes uniquely determine complex human traits and that the solution to these diseases lies in the study of these genes is so prevalent today that it can easily go unnoticed. It is rooted, at least in part, in a powerful myth in the history of science: that Linus Pauling's (1901–1994) discovery of the molecular basis of sickle-cell disease led to therapeutic improvement for patients.[1]

In 1949, Pauling, perhaps the greatest physical chemist of his day, who eventually won two Nobel Prizes, published a paper in

Science entitled "Sickle Cell Anemia, a Molecular Disease." There he showed that the hemoglobin molecules of patients who suffered from the disease, causing severe pain and other symptoms, differed from the hemoglobin molecules of those who did not. More precisely, sickle-cell anemia hemoglobin exhibited a different electric charge, as seen in an electrophoresis apparatus, than did normal hemoglobin. Because the disease was known to be inherited in a Mendelian way, Pauling's result has been taken to mean that a single gene, and the resulting molecule, could determine the occurrence of a disease. Since then, Pauling's paper has been cited almost two thousand times in the scientific literature. More important, it has been included in almost all high school and undergraduate biology textbooks to explain "how human disorders [are] caused by single genes inherited," as one typical textbook put it. For more than half a century, generations of students have learned about the relationship between genes and diseases through this example. After Mendel's peas, Pauling's sickle-cell anemia hemoglobin has been the inescapable story to show how genes (and molecules) determine complex traits.[2]

But Pauling's breakthrough is also used to illustrate a broader point. It is presented as a model of how medical research should be conducted. Beginning in the laboratory, medical research will reveal the true cause of disease, leading to the discovery of new therapeutics to treat patients in the clinic. This model lies at the heart of contemporary biomedicine as a research practice. Since Pauling's work, other examples of medical successes have been used to make the same two points (for example, the genetic disorders PKU and cystic fibrosis), but none has gained the popularity of sickle-cell anemia, which has become a powerful myth and an emblem for a specific research agenda.

Myths, as the French linguist Roland Barthes (1915–1980) put it in his *Mythologies,* are not simply inaccurate statements about the world; they are a specific kind of speech. Myths are a way of collectively expressing something about values, beliefs, and aspirations, even though, taken literally, the content of the myth is not true. As part of the collective memory of every community, myths have an effect on people's identities and destiny. In science, the collective memory of the past shapes research agendas (what questions are worth pursuing) and disciplinary boundaries (who belongs to one discipline or another). Thus, myths not only (imperfectly) reflect the past but also shape the future. For this reason, explaining how and why a myth crystallized in a particular community at a specific time in history is often more illuminating than simply debunking the myth by showing its inaccuracies.[3]

Pauling's Myth about "Molecular Medicine"

The myth constructed around Pauling's discovery focuses on how his molecular approach to the disease "introduced the era of molecular medicine," opening up a successful "rational approach to chemotherapy" and leading to "therapeutic progress" for patients. None of these points is historically accurate, and showing why they are not is revealing about both the history of the life sciences and the history of medicine in the twentieth century.[4]

The precise meaning of "molecular medicine" is rather unclear. But investigations about the relationship between (macro)molecules and diseases were common in the decades before Pauling's paper. The biochemist Frederick Gowland Hopkins's (1861–1947) work on vitamins, for example, established the importance of vitamins in health and disease. A year before Pauling's publication, a hundred-page review described dozens of diseases that had been correlated with quantitative and qualitative alterations of blood proteins examined by electrophoresis, the same technique used by Pauling and his collaborators. Pauling was original in that he was the first to provide "a direct link between the existence of a 'defective' hemoglobin molecule and the pathological consequences of sickle cell disease." For Pauling, the mechanism was rather simple: "the sickle cell anemia hemoglobin molecules might be capable of interacting with one another . . . to cause at least a partial alignment of the molecules within the cell, resulting . . . in the cell's membrane being distorted" and the red blood cells adopting their characteristic sickled shape and the resulting impairment of blood circulation. Thus, Pauling did not introduce "the era of molecular medicine," though he did indeed provide a convincing example of how the etiology (or cause) of a disease could be explained in molecular terms.[5]

Medicine is not only concerned with identifying the causes of diseases; it also aims to treat or, ideally, cure diseases. Behind the claim that Pauling introduced a "rational approach to chemotherapy" lies the criticism that earlier approaches to chemotherapy, such as screening, were "irrational" and unsuccessful. Neither is true. The German physician Paul Ehrlich's (1854–1915) quest for a treatment against syphilis, leading to the first successful chemotherapy, was indeed the result of weeks of tedious testing of chemical compounds until the 606th finally yielded therapeutic benefits. But the choice of the compounds was not irrational. It rested on Ehrlich's theory that if chemical dyes could color cells, it was because they bound chemically to organic matter and thus would constitute good candidates for having a biological effect. The greatest drugs of the twentieth century, from the sulfa drugs to antibiotics to cancer medications, were in large part the result of a similar approach.[6]

The myth is correct in describing Pauling's vision: a detailed molecular understanding of the mechanisms of disease should directly lead to the identification of therapeutic molecules. Immediately after postulating that the interaction between abnormal hemoglobin molecules caused deformation of the red blood cells, Pauling suggested that the chemical that prevented this interaction could cure the disease. It is rarely known that Pauling worked for several years with a physician to test the effect of several molecules on the sickling process. The clinical results all proved negative. The tested molecules did indeed prevent the interaction with hemoglobin, but they also had many other, often toxic, effects. Although Pauling endorsed the view that "man is simply a collection of molecules," the treatment of "man" turned out to be far more complex than he initially envisioned.[7]

Better known is the fact that Pauling considered another way
to eliminate sickle-cell disease: eugenics. In 1968, he went so far
as to suggest that "there should be tattooed on the forehead of
every young person a symbol showing possession of the sickle-
cell gene." A few years earlier, he had argued that the chance that
two parents carrying a mutation causing the sickle-cell disease
would transmit it to their child was far too high (25 percent) "to
let private enterprise in love combined with ignorance take care
of the matter." Except for some limited cases in which manda-
tory policies have been put in place to control for genetic diseases
in populations (such as in Cyprus and Sardinia, where thalas-
semias are common), the focus of therapeutic measure has been
on individuals and their "defective" molecules, following Pauling's initially unsuccessful vision.[8]

One could claim that Pauling was only a bit ahead of his time
and that his views eventually became vindicated. As two researchers at the National Institutes of Health put it in 1995, "it
has taken us almost half a century to cross the threshold from
laboratory to clinic with respect to . . . the 'first molecular disease.'" At last, the authors claimed in the title of their paper,
"Basic Research Reaches the Clinic." But the clinical reality of
sickle-cell anemia is far more nuanced, even today—twenty years
later—than that comment suggests. Indeed, the understanding of
the molecular mechanisms of sickling has led to many insights
about potential therapeutic agents, but to this day none has
reached the clinic. It takes more to transform the clinical reality
of patients than a promising in vitro experiment or an animal
trial. The only chemotherapeutic agent for sickle-cell anemia that
has been brought to the market, hydroxyurea, increases the
production of fetal hemoglobin among adults, thus preventing
the interaction among abnormal hemoglobin molecules, as
Pauling envisioned. Far from constituting a cure, it is part of the
management of pain episodes for sickle-cell patients (a standard
treatment unrelated to Pauling's insight). Furthermore, the de-

velopment of hydroxyurea was not the result of a "rational approach to chemotherapy" based on laboratory research, but of unexpected clinical observations by a pediatrician and epidemiological observations of adult populations who produce fetal hemoglobin.[9]

Occasionally, knowledge of the molecular basis of sickle cell disease has contributed to improvement in therapy. Indeed, newborn screening for sickle-cell anemia is mandated in the United States and elsewhere and is conducted through the identification of abnormal hemoglobin that allows the early diagnosis of sickle-cell disease with great certainty and the beginning of prophylactic treatment for some of the consequences of the disease. But in most parts of the world, diagnosis is carried out through the much simpler and cheaper method of observing sickled blood cells under the microscope, following the method developed by the clinician James B. Herrick (1861–1954) in 1910.[10]

The myth that the discovery of the molecular basis of sickle-cell anemia led to improvements in therapy did not arise by accident. Immediately after the publication of his landmark paper, Pauling went on a crusade to popularize his vision of medical research. In general magazines and scientific journals, as well as at conferences given in the United States, Europe, and Asia, he repeated over and again how the search for molecular abnormalities would eventually lead to cures. In 1952, one newspaper typically carried the title: "Scientist Heralds New Era in Medicine Based on Studies of Molecular Action." Pauling emphasized the broad relevance of his work, claiming that "most cases of mental deficiency can be attributed to molecular diseases." He felt "confident that this knowledge will permit the deduction of improved therapeutic methods." At several conferences, he argued that medicine would be "transformed from its present empirical form into the science of *molecular medicine*." Pauling also attempted (unsuccessfully) to establish a medical research institute at the California Institute of Technology and carried out extensive research (with few results) on the molecular basis of mental diseases. His medical research resonated with his political activism against nuclear fallout, which caused many "molecular diseases."[11]

Pauling's vision about the relationship between laboratory research and clinical applications was embraced by many molecular biologists who were attempting to establish their new discipline in academic institutions. They repeatedly claimed that molecular biology would contribute to medicine, giving them an edge over natural historians and increasing their political appeal. French molecular biologist Jacques Monod (1910–1976), for example, argued in the 1960s that research in molecular biology deserved to be supported because it would allow the emergence of a new field called "molecular pathology." At the same time, a number of these researchers distanced themselves from the clinic, preferring to carry out basic research in the laboratory. Yet the growing number of Nobel Prizes in physiology or medicine awarded to work in molecular biology (in 1962: Francis Crick, James Watson, and Maurice Wilkins; in 1965: François Jacob, André Lwoff, and Jacques Monod; in 1968: Robert W. Holley, Har Gobind Khorana, and Marshall W. Nirenberg; in 1969: Max Delbrück, Alfred Hershey, and Salvador Luria) seemed to vindicate their views of what counted as medical research.[12]

Conclusion

To a large extent, the rise of biomedicine in the twentieth century and its current organization rests on the division of labor and hierarchical relationship between the laboratory and the clinic envisioned by Pauling. But the development of medicine and especially of therapeutics has followed a much more complex path. Only recently have historians started to pay significant attention to the clinical research that goes into developing new therapeutics—not minimizing the importance of laboratory research but challenging that it represents a necessary starting point for medical research.[13] Similarly, Pauling's views about the simple relationship between genes and diseases has driven biomedical research, especially medical genetics, in the second half of the twentieth century. But it has also reinforced one of the most popular myths about the effect of genes on human health, disease, and behavior. As recent studies about human genomes indicate, for most common diseases, it is not one but many genes that are implicated. Furthermore, they don't cause diseases, as the example of sickle-cell disease seemed to indicate; they only increase, and generally just slightly, the risk of occurrence of a disease. The simple world envisioned by Pauling has inspired biomedicine's research agendas. Today, it has largely become a myth that often stands in the way of our understanding of the past, present, and future of health and disease.[14]

THAT THE SOVIET LAUNCH OF *SPUTNIK* CAUSED THE REVAMPING OF AMERICAN SCIENCE EDUCATION

John L. Rudolph

The Soviets' history-making accomplishment—launching a satellite into orbit—created both paranoia and concern that the Soviets had beaten Americans into space. That concern sparked a much-needed revolution in scientific education in the U.S.

—National Public Radio, September 30, 2007

Half a century ago, when the Soviets beat us into space with the launch of a satellite called Sputnik, we had no idea how we would beat them to the moon. The science wasn't even there yet. NASA didn't exist. But after investing in better research and education, we didn't just surpass the Soviets; we unleashed a wave of innovation that created new industries and millions of new jobs.

—President Barack Obama, State of the Union Address, January 25, 2011

The launch of *Sputnik* has long been viewed as a singular, cata-lyzing event in the history of twentieth-century United States of America. In a 2001 book on the subject, freelance nonfiction author Paul Dickson (b. 1939) called it the "Shock of the Century," which is perhaps a fitting appraisal based on the all-caps head-lines that greeted readers opening their newspapers across the country on October 5, 1957, the morning after the launch. President Dwight D. Eisenhower (1890–1969) called the public reaction at the time a "wave of near-hysteria." The orbiting satellite, without question, generated considerable distress nationwide and woke America from its self-satisfied national slumber in the decade of prosperity following World War II. A central element of that awakening was the realization that the country's educational system had slid into mediocrity and was no longer able to keep pace with the challenge posed by Soviet educational training. *Sputnik*'s signals, transmitted at regular intervals overhead, highlighted the flaws in American science education. The Russian satellite, so the story goes, set the country to work fundamentally retooling science teaching in the United States with the aim of ratcheting up disciplinary rigor and scholarly achievement.[1]

Since then, the event has become a cultural touchstone of sorts, marking a time in the popular imagination when an unexpected external threat successfully prodded the country to action. Leaders have pointed to the historical episode time and again to warn against national complacency and spur a collective aspiration for excellence, particularly of the educational variety. When President Barack Obama (b. 1961) recounted the story of *Sputnik* in his 2011 State of the Union speech, he was attempting to rekindle a sense of urgency to launch a new period of science education reform. This time the challenge was coming from the economic sphere of China and India, where advances in science and technology education threatened to move those countries past the United States in the global economic race. With reference to that spark from the past, the president sought to ignite a new wave of innovation and emphasis. But how apt is that metaphor? Did the postwar reforms in science education begin in a flash of activity following *Sputnik*? The historical record on this suggests that they did not.

The retooling of science teaching was well under way prior to the hysteria of *Sputnik*. The most concrete efforts were the National Science Foundation (NSF)-sponsored curriculum projects of the 1950s in the common high school subjects of physics, chemistry, and biology. The first of these was the Physical Science Study Committee (PSSC), a high school physics project organized by Jerrold Zacharias (1905–1986) at the Massachusetts Institute of Technology. PSSC got its start with a grant from the NSF in 1956. But the roots of the project went back to meetings of the Science Advisory Committee in the Office of Defense Mobilization in the administration of President Harry Truman (1884–1972). SAC-ODM, as the group was called, had been established in 1951 (with Zacharias as a member) to advise the government on scientific issues related to national defense. Although the first meetings focused primarily on mundane technical matters, national security officials, lamenting the shortage of scientific manpower, attended them on occasion. With the Korean War in full swing and the government investing millions of dollars in scientific research and development, these officials worried aloud about the country's manpower resources. As Zacharias recalled, "the military would come in and complain that the Russians were getting ahead of us, that we had to do something . . . getting more engineers, more scientists." This prompted him to take on the challenge of remaking high school physics, and in the summer of 1956 he pitched his idea for a new physics curriculum to the education director at the NSF—well over a year before *Sputnik* crossed American skies.[2]

The Colorado-based Biological Sciences Curriculum Study (BSCS), established with funds from the NSF in December 1958, was the second of the science-reform projects. By itself, the timing of BSCS might suggest that *Sputnik* was perhaps responsible for reforming high school biology, if not high school physics. But BSCS, too, had a history that started well before October 4, 1957. As early as 1952, university biologists had begun talking about reorganizing biology teaching in colleges and high schools in meetings of the American Institute of Biological Sciences; in 1954, these same individuals, working under the direction of the National Academy of Science's Division of Biology and Agriculture, set up a committee to begin exploring the reform of biology education. Over three years the committee, supported by funds from both the NSF and the Rockefeller Foundation, worked on developing curricular materials and resources for high school biology classrooms. These were vetted and revised with the assistance of high school and college biology teachers during a summer workshop at Michigan State University in 1957 and were published the following year.[3]

Common to these reform projects (including the first of the chemistry projects, which can be traced back to the summer of 1957 as well) was the supportive financial hand of the NSF; thus, it may be instructive to examine its activities in relation to the *Sputnik* moment. During this time, the NSF itself was a fairly young institution, having been created in 1950. Under its first director, Alan Waterman (1892–1967), the NSF was just feeling its way in the early years. Although Congress had been willing to establish the foundation, it was reluctant to provide much in the way of resources. Using taxpayer dollars for "pure research" that might not have any practical payoff (which was the NSF's primary mission) was never an easy sell on Capitol Hill. Thus, to stay in the good graces of Congress, Waterman avoided ventures that might invite scrutiny or controversy. This was especially true when it came to education, a topic rife with controversy in the early 1950s.[4]

Politicians, educators, and administration officials frequently clashed over a range of school issues. At the heart of these was the question of what role the federal government should assume in an enterprise long controlled by local districts. Many local officials and teacher groups called for federal money to ease the crushing burden of skyrocketing enrollments brought on by the baby boom, which created a need for thousands of new buildings and more teachers to staff them. But providing relief was no easy matter. Federal involvement in local school affairs was complicated by segregation practices in the South and the equally pressing needs of parochial schools. Federal dollars would likely come with strings attached, and those strings, many believed, had the potential to pull apart many long-standing local customs. Issues of religion and race in particular had repeatedly stalled congressional efforts to provide aid.[5]

Nevertheless, Congress had devoted a small but explicit part of the NSF's mission to promote education in the sciences. Given the political minefield that lay before it, the NSF limited its education offerings to colleges, steering clear of direct involvement with the lower schools. At the college level, the NSF funneled the lion's share of its educational resources to graduate fellowships for scientists in training. Officials in the education division, however, recognized the need to promote better teaching at the beginning of the pipeline. In the summer of 1952, NSF officials began planning a small summer-institute program that would bring high school science teachers up to date in the latest disciplinary content in their fields. The idea was to target practicing teachers directly—during their summers off—in order to avoid becoming entangled in school-funding or curriculum issues. These summer institutes were modeled after a series of summer teacher-training programs offered as early as 1945 by General Electric, which had been a leader in seeking to improve science teaching. These efforts reflected the scientific manpower concerns voiced perennially throughout the later 1940s and 1950s.[6]

The NSF's deliberately cautious approach to reforming science education, however, was cast aside rather dramatically in 1956, when Congress suddenly increased its education budget to over $10 million in one year—an eightfold increase from the previous year. The funds were allocated in an effort to push the agency to support more radical reform. But it wasn't the launch of *Sputnik* that prompted this extraordinary surge in funding. It was, rather, the publication in the previous year of a somewhat modest report on the Soviet educational system prepared by a young doctoral student at Harvard University's Russian Research Center, Nicholas DeWitt (1923–1995).[7]

DeWitt's report, *Soviet Professional Manpower: Its Education, Training, and Supply,* was the culmination of research into this topic that DeWitt had begun as early as the spring of 1952. When it came out in the summer of 1955, it quickly caught the attention of Texas representative Albert Thomas (1898–1966), who chaired the House Appropriations Subcommittee for Independent Offices (the committee that controlled the NSF's budget). The subcommittee members, like many other U.S. politicians, had long been skeptical of federal involvement in education. However, DeWitt's report, perhaps combined with the fact that the Soviet Union had successfully detonated a deliverable hydrogen bomb in November 1955, seemed to turn the tide. At the House subcommittee hearings in January 1956, Thomas acknowledged the source of his conversion: "This little book, *Soviet Professional Manpower,* I read word for word . . . and after reading it I completely reversed my thinking." Russia's focus on high school science teaching, he exclaimed, "is the most alarming thing that I can imagine. . . . Lord help us if they ever reach the point where they are ahead of us." Following these hearings, the NSF found itself with $10.9 million to spend on nonfellowship education programs. The next year *Sputnik* arrived, which was followed by the National Defense Education Act in 1958.[8]

One might argue that these federal-agency and scientist-led efforts to remake American science education existed behind the scenes and were only brought to light and embraced by the public with the shock of the *Sputnik* launch. But even this overstates the influence of the Russian satellite. There was nothing "behind the scenes" about the drumbeat of criticism the schools endured after the Second World War. The late 1940s and early 1950s saw repeated, scathing critiques of public education, which came out alongside annual media reports of teaching shortages, overflowing school buildings, and inadequate facilities. Every year, it seemed, the crisis got worse. High-profile proposals for reforming science teaching appeared as early as 1945, when the Harvard Red Book report, *General Education in a Free Society,* was released;

curricular reforms were pushed again at the White House Conference on Education ten years later. By the fall of 1957, the fire for changes in science education was already burning steadily; reform was inevitable. *Sputnik*'s appearance in American skies certainly added fuel to that fire, but contrary to myth, it wasn't the spark that started it.[9]

IV

GENERALIZATIONS

THAT RELIGION HAS TYPICALLY IMPEDED

THE PROGRESS OF SCIENCE

Peter Harrison

Science and faith are fundamentally *incompatible,* and for precisely the same reason that irrationality and rationality are incompatible. They are different forms of inquiry, with only one, science, equipped to find real truth. . . . And *any* progress—not just scientific progress—is easier when we're not yoked to religious dogma.

　—Jerry A. Coyne, "Science and Religion Aren't Friends" (2010)

The conflict between religion and science is inherent and (very nearly) zero-sum. The success of science often comes at the expense of religious dogma; the maintenance of religious dogma always comes at the expense of science.

　—Sam Harris, "Science Must Destroy Religion" (2006)

One of the most widespread misconceptions about science concerns its historical relationship to religion. According to this pervasive myth, these two enterprises are polar opposites that compete to occupy the same explanatory territory. The history of Western thought is understood in terms of a protracted struggle between these opposing forces, with religion gradually being forced to yield more and more ground to an advancing science that offers superior explanations. Wherever possible, religion has resisted this ceding of territory, thus hindering the advance of science. While historians of science have long ago abandoned this simplistic narrative, the "conflict myth" has proven to be remarkably resistant to their demythologizing efforts and remains a central feature of common understandings of the identity of modern science.

Public rehearsals of this myth are typically prompted by contemporary instances of religiously motivated rejections of science—most often episodes of antievolutionary sentiment. In these discussions, the myth is deployed to explain why we should not be surprised at present outbreaks of science–religion conflict, since these are simply the contemporary manifestation of a long-standing historical pattern in which science has always been resisted by religion. It is assumed that these two entities, "science" and "religion," will inevitably clash, owing to their inherent nature. Religion is said to be based on authority, ancient religious texts, blind faith, or simply irrational prejudice; science, on reason and common sense. Conflict arises out of the fact that these two enterprises seek to offer explanations for the same things but from these incompatible starting points.

Crucial to the argument are the historical events thought to be emblematic of this antagonistic relationship. The favored examples are the 1633 condemnation of Galileo Galilei (1564–1642) and the reception of Charles Darwin's (1809–1882) *Origin of Species* (1859), but there is also a long list of supporting characters and episodes: the hostility of Tertullian (ca. 160–225), the father of Latin Christianity, to Greek philosophy; the murder of mathematician and philosopher Hypatia (ca. 350–415) at the hands of a Christian mob; Pope Callixtus III's (1378–1458) excommunication of a comet in 1475; medieval belief in a flat earth (see Myth 2); the church's banning of dissection; resistance to Copernicanism because it would "demote" human beings from the center of the cosmos (see Myth 3); Giordano Bruno's (1548–1600) execution in 1600 as a martyr to science; and religious opposition to medical advances, such as inoculation and anesthesia. These historical examples serve the purpose of not merely illustrating the general likelihood of religious resistance to science but also implying that such resistance really belongs to some dim and distant "dark ages"(see Myth 1). In recent years, historians of science have conclusively shown that there is little historical basis for the myth. Much of the adduced evidence is simply false—including the excommunication of the comet, belief in the flat earth, the banning of dissection, resistance to Copernicanism because of its assault on human dignity, Bruno's execution on scientific grounds, and resistance to inoculation and anesthesia.[1]

Other episodes have a firm historical basis but are far more complicated than simple instances of a science–religion conflict. In the case of Galileo, it is indisputable that he was condemned by the Inquisition in 1633 for holding and teaching Copernican views—that is, that the earth was in motion around the sun. However, this was by no means a typical response of the Catholic Church to "science" in general. At the time, the church was the major sponsor of astronomical research.[2] Moreover, the relevant science was by no means clear-cut, with scientific authorities divided on the relative merits of competing cosmological systems. Thus, Galileo enjoyed support from within the church as well as opposition from the scientific establishment. Neither was he tortured or imprisoned. So while the facts of Galileo's condemnation are not in dispute, that they were typical of a Catholic attitude toward science, or that the episode was primarily about "science vs. religion," is highly questionable. In the case of Darwin, similar considerations apply. Although there was, undoubtedly, religious resistance to the idea of evolution by natural selection, Darwinism had both significant religious supporters and influential scientific detractors.[3] As for the phenomenon of "scientific creationism," which is now the most conspicuous manifestation of religiously motivated resistance to evolution, this is essentially a twentieth-century development and was not a feature of initial reactions to Darwin's theories.[4]

On the other side of the ledger, historians have also drawn attention to the ways in which religious considerations have played a positive role in the content and conduct of the sciences. The medieval universities, which were the chief sites of scientific activity in the later middle ages, were founded and supported by the Catholic Church.[5] Key seventeenth-century figures—such as Johannes Kepler (1571–1630), Robert Boyle (1627–1691), Isaac Newton (1643–1727), and John Ray (1627–1705)—were clearly motivated by religious considerations and said as much. The idea of laws of nature, fundamental to the conduct of modern physics, was originally a theological conception. Arguably, experimental method, too, owes much to theological conceptions of human nature that stress the fallibility of our cognitive and sensory capacities. More generally, it has been argued that Protestantism promoted a desacralization of nature, providing a hospitable environment for the development of modern science. Lastly, religion has been important for establishing the social legitimacy of science, owing to the identification of science as a means of redemption and a form of "priestly" activity, and to the strong partnership between natural theology and the natural sciences that characterized science in England from the seventeenth to the nineteenth centuries.[6] It follows that even if the Galileo affair or the religious reception of Darwinism could be regarded as clear-cut examples of religion impeding science, they could not be said to exemplify any general pattern or essential relationship.

A further difficulty with the myth is its tendency to regard science and religion as monolithic and unchanging historical forces, or successive epochs. In fact, for long periods of Western history, scientific and religious concerns overlapped considerably and in ways that make it difficult to distinguish between science and religion as we now understand them. Indeed, in English-speaking countries, no one spoke about science and religion until the nineteenth century. So the idea of *any* ongoing historical relationship between science and religion calls for the imposition of modern categories onto past actors for whom the distinction would have been essentially meaningless.[7]

Given the flimsy historical and conceptual foundations of the conflict myth, it is natural to ask where it originated and why—in spite of the best efforts of historians—it persists. Historians of science have typically traced the origins of the myth to two nineteenth-century works—John William Draper's (1811–1882) *History of the Conflict between Religion and Science* (1874) and Andrew Dickson White's (1832–1918) *History of the Warfare of Science with Theology in Christendom* (1896). But the myth well predates them. Versions of it can already be found in seventeenth-century Protestant polemics against Catholicism that sought to align papism with ignorance, superstition, and resistance to new knowledge. The influential Puritan minister and fellow of the Royal Society, Cotton Mather (1663–1728), contended that Europe had been plunged into darkness during the Catholic Middle Ages and that the revival of letters and reformation of religion had together paved the way for "the advances of the sciences."[8] The treatment of Galileo at the hands of the Inquisition was a special gift to Protestant apologists, who pioneered the use of the Galileo affair for propaganda purposes. In 1638, the poet John Milton (1608–1674) visited Galileo, then under house imprisonment in Florence, and used the occasion to reflect on the contrast between the philosophical freedom of Protestant England and the scientific censorship of Catholic Italy.[9]

This "Protestant position" would subsequently be adopted and applied more generally by French *philosophes* and key Enlightenment figures. Voltaire (1694–1778) observed that had Isaac Newton been born in a Catholic country and not Protestant England, rather than becoming a scientific celebrity he might well have found himself clothed in the robes of a penitent and burnt at an auto-da-fé.[10] Jean d'Alembert (1717–1783) wrote of Galileo's troubles in that great monument of the Enlightenment, the *Encyclopédie,* concluding that poorly informed theologians had habitually waged "open war" against philosophy (that is, science).[11] The Enlightenment idea of progress was thus difficult to disentangle from an accompanying narrative about the stultifying effects of religion on the advance of knowledge. Summing up this view of history, Nicolas de Condorcet (1743–1794) announced in his *Sketch for a Historical Picture of the Progress of the Human*

Spirit (1795) that "the triumph of Christianity was the signal for the complete decadence of philosophy and the sciences."[12]

This version of events came to be incorporated, in turn, into the grand progressive models of history that were common in the nineteenth century. Best known, perhaps, is Auguste Comte's (1798–1857) simplistic but appealing idea that history flows through three successive stages: theological, metaphysical, and positive (scientific). The idea that religion represents a backward phase of human development, destined to be overtaken by a scientific one, became commonplace in social-scientific understandings of human development, as well as in general histories. In England, historian Henry Thomas Buckle (1821–1862) spread the message that the process of civilization was hastened by scientific skepticism and impeded by the credulous conservatism that characterized religion. Meanwhile, in Germany, Friedrich Lange (1828–1875), in his influential *History of Materialism* (1866), wrote of how "every system of philosophy entered upon an inevitable struggle with the theology of its time."[13]

Given this background of progressive understandings of history that had already attributed a special, inhibitory role to religion, by the time Draper and White took up their pens, all that remained was to fill in the blanks. This they managed with considerable alacrity, combining ingenuity and imagination to compile the now-standard catalog of historical episodes illustrative of an enduring conflict.[14] Ironically, then, what began as a single-purpose weapon in the arsenal of Protestant apologists—a role still evident in Draper's largely anti-Catholic tome—became a blunderbuss to be deployed indiscriminately against all forms of religion.

Why, given its fragile foundation in reality, does this somewhat old-fashioned myth persist? There are a number of reasons. For a start, there are conspicuous contemporary instances of religious resistance to science—most obviously, the rejection of evolutionary theory by scientific creationists. This apparently indisputable instance of science–religion conflict continues to fuel the myth on the assumption that the present must resemble the

past. It should be pointed out in this context that antievolutionists actually tend to be pro-science in general terms and, for
this reason, couch their religious beliefs in scientific language; it
is just that they oppose one particular scientific theory and have
routinely challenged its scientific standing. Indeed, one of the
more interesting findings of the World Values Survey is that dis-
trust of science is greatest in those countries that are the most
secularized, and least in those that are most religious.[15] That notwithstanding, it is clear that the conflict myth most often gets an
airing in responses to the activities of antievolutionists.

Related to this is a more general fear associated with the so-
called return of religion. In the middle of the twentieth century,
the prevailing vision of history predicted a largely religion-free
future in which a secular, scientific worldview would become the
default position. For reasons well known, this failed to obtain.
Present concerns about religious fundamentalism, and militant
Islam in particular, give new normative force to the conflict myth.
Science is perceived by some to be the vehicle of a form of secular
enlightenment. The conflict between science and religion is thus
more than an abstract description of a distant past: it has become
the founding myth of a crusade to secure a threatened secular
future. This sentiment clearly informs the views of Jerry Coyne
(b. 1949) and Sam Harris (b. 1967), noted at the outset. As Harris
plainly expresses it: there is an ongoing conflict between science
and religion, and science must win it.

But at the most general level, the conflict myth has something
for everybody. Its irresistible appeal lies in the various plotlines
that pit the lone genius against the faceless men or expose the
apparent idiocies of inflexible institutions. Ultimately, it suggests
the triumph of reason over superstition, of good over evil. This is
a comforting and congenial myth that also assures us of our cultural and intellectual superiority. In spite of the evidence against
it, while it continues to fulfill these functions, it is difficult to see
it disappearing any time soon.

THAT SCIENCE HAS BEEN LARGELY A SOLITARY ENTERPRISE

Kathryn M. Olesko

I know not what I may seem to the world; but to myself, I seem to have been only like a boy, playing on the sea-shore, and diverting myself, in now and then finding a smoother pebble or a prettier shell than ordinary, whilst the great ocean of truth lay all undiscovered before me.

—Isaac Newton, as reported by Edmond Turnor (1806)

It's difficult to imagine how I would ever have enough peace and quiet in the present sort of climate to do what I did in 1964.

—Peter Higgs, as quoted in *The Guardian* (2013)

Three hundred years separate Isaac Newton (1643–1727) and Peter Higgs (b. 1929), but they share in common a belief in the value of solitude for scientific work. The image of the scientist as the solitary genius who works alone for hours on end, emotionally separated from friends and family and oblivious to the humdrum needs of daily life, is deeply embedded in culture. Around 1800, William Blake's iconic and well-known depiction of Newton as the godlike geometer, united with his natural surroundings and deep in contemplation on the shores of the sea, epitomized the solitude of scientific creativity. Popular culture later transformed the image, but the element of solitude remained intact. In contrast to Blake's divine Newton, Mary Shelley created the lonely and mad Dr. Victor Frankenstein (1818); H. G. Wells, the sinister and isolated Dr. Moreau (1896); and director Roland Emmerich, the unbalanced and slightly wild Dr. Brackish Okun in the movie *Independence Day* (1996).

Mid-twentieth-century biographical studies of Isaac Newton provide one of the most well-known examples of the myth of the solitary genius. Forced to flee Cambridge, where the plague had spread, and take refuge on his mother's farm in Lincolnshire, Newton was reputed to have ventured into a garden where a falling apple sparked the idea of universal gravitation in a moment of divine inspiration (see Myth 6). Scholars have since dubbed 1665–1666 Newton's "miraculous year," the year in which he discovered gravity, the composition of white light, and calculus. After the Lincolnshire years, Newton's solitude, self-neglect, detachment from society, and tendency to become lost in thought became legendary. Although educated at Cambridge, he presented himself as someone who had no need for pedagogy, schooling, and the social processes by which knowledge was learned. He became the self-taught genius who had no teacher—the perfect solitary scholar. So engrained did this image become that it did not matter whether or not Newton uttered the words in the epigraph to this essay (he probably did not). Alexander Pope's (1688–1744) proposed epitaph for Newton's grave expressed the spirit of this solitary genius:

> Nature and Nature's Laws lay hid in Night.
> *God said*, Let Newton be! *and All was* Light.[1]

It is not at all surprising, then, that Newton's legendary work habits dictated the title of a biography: Newton was "never at rest."[2]

Why has the myth of science as a largely solitary enterprise endured? Like all myths, it is a story that legitimates aspects of the social, cultural, economic, or political order. A transcendent engagement with nature in solitude evokes scenes of religious revelation: St. John in the desert, St. Jerome in his study, Jesus in the garden of Gethsemane. The ideals of liberal individualism and

Western rationality are embedded in solitary creativity. The myth presumes that detachment and isolation are necessary preconditions for objectivity, and so for truth. In the Western tradition of liberalism following John Stuart Mill (1806–1873), the voice in the wilderness addresses a reality others cannot see and so must be heeded. In the context of these traditions, the scientist in solitude is a secular saint: ascetic, self-denying, and, above all, self-disciplined. No wonder, then, that exceptional scientists, like saints, have "miraculous years" (Albert Einstein's was 1905; see Myth 18). Emotionality and emotional attachments—especially to family and loved ones—are unnecessary and potentially dangerous distractions corrupting the heroic search for the secrets of nature. This myth is thus a self-validating ideology. Powerful as these associations are, though, ideology accounts for only part of its persistent appeal.[3]

Science itself is partly to blame. The reward systems of science celebrate the individual. Scientists are honored in perpetuity by eponymous laws and constants and entities—such as Newton's laws, Mendel's Laws, Planck's constant, or the Higgs boson—which project scientific discoveries as individual achievements. Nobel Prizes are awarded to individuals, not groups or teams. Scientific textbooks credit individual scientists with discovery and invention. Evidence for the myth is nearly everywhere in scientific culture, and it is difficult to overcome.

The myth persists, though, less because of science than because stories about it have been told in particular ways. The culprit is history: as scientists tell it, as historians write it, and as students understand it. History as rendered in scientific textbooks is mostly about individuals, not scientists working in teams or in communication with one another. So physicist David Park told the story of the wave nature of light through a series of individual discoveries, from Thomas Young (1773–1829) in 1802 to James Clerk Maxwell (1831–1879) in 1861. He left out, interestingly, the most politically contentious contributor, Augustin Fresnel (1788–1827), and his work on diffraction. The omission is telling. Fresnel's contact with the British community and the politics of science, especially in France's Academy of Sciences, of deciding whether light was composed of waves or particles played a powerful role in shaping the fate of the wave theory of light.[4] Social systems and political intrigue simply did not fit Park's conception of history.

Textbook entries that seem to confirm science as a solitary enterprise have a ripple effect on history. Consider the history of the element yttrium and its textbook-designated discoverer, the relatively unknown Finnish chemist Johan Gadolin (1760–1852). Gadolin never claimed to have discovered a new earth in 1794— only that he could not identify part of the composition of a black stone he had found in the Ytterby quarry in Finland. When other chemists scrutinized his results over the following decades, they named the new earth yttria. The act of naming consolidated history: Gadolin became the most important contributor to the discovery of yttria even though other investigators did more to uncover its properties. When chemical textbooks reclassified the earths as elements in the nineteenth century, each element acquired a story of its origin. Complex stories eventually evolved into simpler ones in scientific handbooks, which in the end identified Gadolin as the *only* discoverer of yttrium. Popular histories of science in the early twentieth century accepted the party line.[5]

In each retelling of yttrium's discovery—from textbooks to handbooks to popular expositions—Gadolin's role expanded while that of the scientific community waned. This literary process exemplifies one way the myth of science as a solitary enterprise could be perpetuated within and beyond the scientific community. But scientists are not the only ones responsible for the myth's grip on history.

Until well into the twentieth century, historians were also responsible for endorsing it. Conventions of historical writing help to explain why. Educational psychologists argue that precollege students tend to view history in terms of individual agency rather than as the unfolding of larger processes. So the discovery of

America in 1492 is about Christopher Columbus, King Ferdinand, and Queen Isabella, rather than about the Kingdom of Castile undergoing social, religious, and economic change, and engaging in commercial competition. This approach is commonly known as "Great Man" history, and many of the early classical narratives in the history of science adopted it. Histories of the Scientific Revolution—the period from Nicolaus Copernicus's (1473–1543) proposal that the sun replace earth as the center of the universe in 1543 to shortly after the publication of Isaac Newton's theory of universal gravitation in 1687—were and still are addicted to narratives guided largely by individuals and their discoveries.[6]

Concurrent with the social movements of the 1960s and later—civil rights, the women's movement, and antiwar protests—scholars revolutionized the historical study of the sciences. Although varied, these new approaches shared a common belief in the inherently social nature of scientific practice. Most fall under the rubric of social constructivism. Constructivism emphasizes context, community, controversy, communication, and many other social dimensions of scientific practice, and it blurs the boundary between science and society. Scientific pedagogy, for instance, which used to be viewed as part of institutional history, became a crucial site of generational reproduction and knowledge formation and transmission. At an earlier time, facts were discovered. From the constructivist's perspective, facts are not born—they are made. The creation of knowledge, once regarded as an individual pursuit, became a collective enterprise. Scientific discovery, formerly the final stage of knowledge creation, now started as a local phenomenon. Only gradually and through a communal process did the results of discovery become universal. Moreover, the larger contexts of scientific practice—social, political, economic, and cultural—could insert themselves into the production of knowledge at any point. The assumption undergirding constructivism is that science and society are inseparable. To know one, you had to know the other.[7]

There was no room in constructivism for the myth of science as a solitary enterprise. The establishment of scientific societies in the seventeenth century, so the argument now goes, was partly to neutralize any chance that scientific practice would resemble religious practice, in which solitude was condoned. Those who claimed that solitude was a precondition for scientific discovery were merely speaking rhetorically. A leading practitioner of constructivism concluded: "The solitary philosopher is therefore only a man imitating God."[8]

From a constructivist perspective, then, a figure such as Newton looks very different. How solitary was he as a scientific practitioner? He certainly never ventured very far from Cambridge, London, and Lincolnshire. And he never visited a beach. Yet despite his relatively stationary existence, he was wired into a vast global network of numerical data on tidal levels, the length of pendulums, and the position of comets, on which he drew to support empirically his theory of universal gravitation. This information could only have reached the shores of England through connections established by trading companies, Jesuit missionaries, astronomers, and the correspondence network of scholars known as the Republic of Letters. Natural philosophers, astronomers, mariners, dockyard workers, and traders shipped their local data to Newton, or Newton sent his emissaries to distant ports and sites to gather the quantitative information that he needed. This relay of information was never simply a handoff because both the data and their producers had to be assessed and analyzed for reliability—a task Newton mastered by developing the means to reconcile discordant data by taking their averages. Thus, Newton's natural philosophy would not have been possible without Great Britain's commercial revolution and the global trade network of which it was a part. Similarly, Charles Darwin's (1809–1882) evidence of biological evolution came from sources embedded in Britain's imperial network. Stories of Newton or Darwin as solitary geniuses are cultural constructions that ignore the important role of context in their lives.[9]

Like all myths, the myth of the solitary scientist distorts the past. It does so not only by sins of commission but also by sins of omission. Its corollary is the myth that science is a (white) masculine pursuit. Both myths render invisible the contributions of women, people of color, and technicians. Female scientists, assistants, calculators, and technicians who for a long time worked alone or nearly alone were overshadowed by their male counterparts. Notwithstanding the controversy over whether or not James Watson (b. 1928) and Francis Crick (1916–2004) used Rosalind Franklin's (1920–1958) X-ray diffraction images with her explicit permission in their 1953 discovery of the structure of DNA, her contribution was crucial to the discovery—and yet she received less credit for it than Watson and Crick did. In a similar vein, when Cecilia Gaposchkin (1900–1979) discovered that hydrogen was the most abundant element in the universe in 1925, male astronomers remained unconvinced until Henry Norris Russell (1877–1957), director of the Princeton University Observatory, published the result and cited Gaposchkin in his footnotes.[10]

The newer approaches to the history of science have resulted in a growth of interest in the roles played by women and other formerly neglected groups in the scientific enterprise. The historian's broader vision of the demographic base of practicing scientists cannot sustain a conception of the scientific persona whose traits were only masculine. More closely considering the role of women illuminates how compatible family life and the pursuit of science could be. Combining the two was sometimes even necessary in sciences such as astronomy, where women aided nocturnal celestial observations during certain eras. The history of women in science thus resulted in some of the most important correctives to the myth of science as a solitary enterprise.[11]

Peter Higgs's statement in the epigraph was an admission that the myth of science as a solitary enterprise persisted into the twentieth century. But it also acknowledged that in the twenty-first century, science could not be practiced alone—and in quietude.

Thoughts produced in solitude do not become part of scientific knowledge unless subjected to a tightly controlled peer-review process and then offered to the scientific community for further discussion, debate, and review at meetings and conferences. Collaborative researchers in a laboratory setting have evolved into international teams of researchers who combine their data in multiauthored publications. The Intergovernmental Panel on Climate Change, for instance, has hundreds of authors and dozens of editors who vote on every decision in its reports.[12] The epistemological space of science operates in a democratic fashion. No more solitary than life itself, science is fundamentally and deeply a collaborative and social enterprise.

THAT THE SCIENTIFIC METHOD ACCURATELY
REFLECTS WHAT SCIENTISTS ACTUALLY DO

Daniel P. Thurs

The scientific method is the process by which science is carried
out.

— *Wikipedia,* "Scientific Method"

It's probably best to get the bad news out of the way first. The
so-called scientific method is a myth.[1] That is not to say that scientists don't do things that can be described and are unique to
their fields of study. But to squeeze a diverse set of practices that
span cultural anthropology, paleobotany, and theoretical physics
into a handful of steps is an inevitable distortion and, to be blunt,
displays a serious poverty of imagination. Easy to grasp, pocket-
guide versions of the scientific method usually reduce to critical
thinking, checking facts, or letting "nature speak for itself," none
of which is really all that uniquely scientific. If typical formulations were accurate, the only location true science would be taking
place in would be grade-school classrooms.

Scratch the surface of the scientific method and the messiness spills out. Even simplistic versions vary from three steps to
eleven. Some start with hypothesis, others with observation. Some
include imagination. Others confine themselves to facts. Question a simple linear recipe and the real fun begins. A website
called Understanding Science offers an "interactive representation" of the scientific method that at first looks familiar. It in-

cludes circles labeled "Exploration and Discovery" and "Testing Ideas." But there are others named "Benefits and Outcomes" and "Community Analysis and Feedback," both rare birds in the world of the scientific method. To make matters worse, arrows point every which way. Mouse over each circle and you find another flowchart with multiple categories and a tangle of additional arrows.[2]

It's also telling where invocations of the scientific method usually appear. A broadly conceived method receives virtually no attention in scientific papers or specialized postsecondary scientific training. The more "internal" a discussion—that is, the more insulated from nonscientists—the more likely it is to involve procedures, protocols, or techniques of interest to close colleagues.[3] Meanwhile, the notion of a heavily abstracted scientific method has pulled public discussion of science into its orbit, like a rhetorical black hole. Educators, scientists, advertisers, popularizers, and journalists have all appealed to it.[4] Its invocation has become routine in debates about topics that draw lay attention, from global warming to intelligent design. Standard formulations of the scientific method are important only insofar as nonscientists believe in them.

Now for the good news. The scientific method is nothing but a piece of rhetoric. Granted, that may not appear to be good news at first, but it actually is. The scientific method as rhetoric is far more complex, interesting, and revealing than it is as a direct reflection of the ways scientists work. Rhetoric is not just words; rather, "just" words are powerful tools to help shape perception, manage the flow of resources and authority, and make certain kinds of actions or beliefs possible or impossible. That's particularly true of what Raymond Williams called "keywords." A list of modern-day keywords include "family," "race," "freedom," and "science." Such words are familiar, repeated again and again until it seems that everyone must know what they mean. At the same time, scratch their surface, and their meanings become full of messiness, variation, and contradiction.[5]

Sound familiar? Scientific method is a keyword (or phrase) that has helped generations of people make sense of what science was, even if there was no clear agreement about its precise meaning—*especially* if there was no clear agreement about its precise meaning. The term could roll off the tongue and be met by heads nodding in knowing assent, and yet there could be a different conception within each mind. As long as no one asked too many questions, the flexibility of the term could be a force of cohesion and a tool for inspiring action among groups. A word with too exact a definition is brittle; its use will be limited to specific circumstances. A word too loosely defined will create confusion and appear to say nothing. A word balanced just so between precision and vagueness can change the world.

This has been true of the scientific method for some time. As early as 1874, British economist Stanley Jevons (1835–1882) commented in his widely noted *Principles of Science,* "Physicists speak familiarly of scientific method, but they could not readily describe what they mean by that expression." Half a century later, sociologist Stuart Rice (1889–1969) attempted an "inductive examination" of the definitions of the scientific method offered in social scientific literature. Ultimately, he complained about its "futility." "The number of items in such an enumeration," he wrote, "would be infinitely large."[6]

And yet the wide variation in possible meanings has made the scientific method a valuable rhetorical resource. Methodological pictures painted by practicing scientists have often been tailored to support their own position and undercut that of their adversaries, even if inconsistency results.[7] As rhetoric, the scientific method has performed at least three functions: it has been a tool of boundary work, a bridge between the scientific and lay worlds, and a brand that represents science itself. It has typically fulfilled all these roles at once, but they also represent a rough chronology of its use. Early in the term's history, the focus was on enforcing boundaries around scientific ideas and practices. Later, it was used more forcefully to show nonscientists how science could be made relevant. More or less coincidentally, its invocation assuaged any doubts that real science was present.

Timing is a crucial factor in understanding the scientific method. Discussion of the best methodology with which to approach the study of nature goes back to the ancient Greeks. Method also appeared as an important concern for natural philosophers during the Islamic and European Middle Ages, whereas many historians have seen the methodological shifts associated with the Scientific Revolution as crucial to the creation of modern science.[8] Given all that, it's even more remarkable that "scientific method" was rarely used before the mid-nineteenth century among English speakers, and only grew to widespread public prominence from the late nineteenth to the early twentieth centuries, peaking somewhere between the 1920s and 1940s (see Figures 26.1 and 26.2).[9] In short, the scientific method is a relatively recent invention.

But it was not alone. Such now-familiar pieces of rhetoric as "science and religion," "scientist," and "pseudoscience" grew in prominence over the same period of time.[10] In that sense, "scientific method" was part of what we might call a rhetorical package, a collection of important keywords that helped to make science

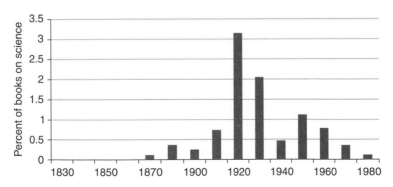

Figure 26.1. Percent of all books published, by decade, with the phrase "scientific method" in the title. (Data source: Library of Congress.)

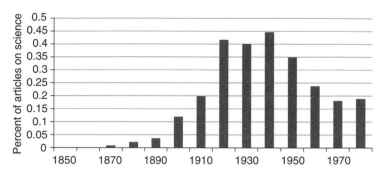

Figure 26.2. Percent of all magazine articles with the phrase "scientific method" in the title. (Data source: Periodicals Contents Index.)

comprehensible, to clarify its differences with other realms of thought, and to distinguish its devotees from other people. All of this paralleled a shift in popular notions of science from general systematized knowledge during the early 1800s to a special and unique sort of information by the early 1900s. These notions eclipsed habits of talk about the scientific method that opened the door to attestations of the authority of science in contrast with other human activities.[11]

Such labor is the essence of what Thomas Gieryn (b. 1950) has called "boundary-work"—that is, exploiting variations and even apparent contradictions in potential definitions of science to enhance one's own access to social and material resources while denying such benefits to others.[12] During the late 1800s, the majority of public boundary-work around science was related to the raging debate over biological evolution and the emerging fault line between science and religion. Given that, we might expect the scientific method to have been a prominent weapon for the advocates of evolutionary ideas, such as John Tyndall (1820–1893) or Thomas Henry Huxley (1825–1895). But that wasn't the case. The notion of a uniquely scientific methodology was still too new and lacked the rhetorical flexibility that made it useful. Instead, the loudest invocations of the scientific method were by those who hoped to *limit* the reach of science. An author in a magazine called *Ladies' Repository* (1868) reflected that "every generation, as it accumulated fresh illustrations of the scientific method, is more and more embarrassed at how to piece them in with that far grander and nobler personal discipline of the soul which hears in every circumstance of life some new word of command from the living God."[13]

By the twentieth century, references to "scientific method" had become a common element of public discussion. The term had accumulated a variety of meanings that allowed it to become a useful rhetorical tool. Meanwhile, the actual content of science seemed to be receding behind increasingly technical barriers. In 1906, a columnist in the *Nation* lamented the greater complexity of scientific knowledge. "One may say," the author observed, "not that the average cultivated man has given up on science, but that science has given up on him."[14] The scientific method remained the only stable bridge to make what happened in the lab relevant to the realm of ordinary life. It showed why science was important and provided an outlet for harnessing that importance, one open even to the average citizen otherwise bewildered by scientific information.

Under such conditions, it was no wonder when some people asserted that the "greatest gift of science is the scientific method."[15] In his 1932 address to journalists in Washington, D.C., physicist Robert Millikan (1868–1953) informed his audience that the "main thing that the popularization of science can contribute to the progress of the world consists in the spreading of a knowledge of the method of science to the man in the street."[16] Educators especially promoted the scientific method as a way of bringing science into the classroom.[17] Before the educational section of the American Association for the Advancement of Science in 1910, John Dewey (1859–1952) charged that "science has been taught too much as an accumulation of ready-made material with which students are to be made familiar and not enough as a method of thinking." In 1947, the *47th Yearbook of the National Society for the Study of Education* declared that there "have been few points in educational discussions on which there has been greater agreement than that of the desirability of teaching the scientific method."[18]

As science became a more powerful force in modern society and culture, thanks in part to invocations of the scientific method, growing numbers sought to take advantage of its prestige. This was especially important for social scientists, who were often seen as scientific pretenders. John B. Watson (1878–1958), the central figure in the behaviorist program, agreed in 1926 that psychology's methods "must be the methods of science in general." That same year, the Social Science Research Council retooled one of its subgroups into the Committee on Scientific Method. A conference held under its auspices eventually generated the massive *Methods in Social Science*.[19] Journalists who looked to social science as a guide during the 1920s and 1930s also turned to the scientific method. In 1928, George Gallup (1901–1984), the founder of the Gallup poll, completed a dissertation at the University of Iowa on "An Objective Method for Determining Reader-Interest." Two years later, he presented an article called "A Scientific Method for Determining Reader-Interest." In both cases, he advocated examining newspapers along with readers, noting their reactions.[20]

During the early 1900s, references to scientific medicine, scientific engineering, scientific management, scientific advertising, and scientific motherhood all spread, often justified by adoption of the scientific method. Amid the spread of totalitarianism in the 1930s and 1940s, the ability of the scientific method to sustain a balance between an open and a critical mind foreshadowed a true "science of democracy."[21] Consumers in a new, advertising-driven marketplace encountered less high-minded examples in books such as *Eby's Complete Scientific Method for Saxophone* (1922), Martin Henry Fenton's *Scientific Method of Raising Jumbo Bull-frogs* (1932), and Arnold Ehret's *A Scientific Method of Eating Your Way to Health* (1922). Eby, for one, never spelled out his complete scientific method. But he didn't need to. Like the swoosh on a Nike shoe, the scientific method only needed to be displayed on the surface.

After the middle of the twentieth century, the scientific method continued to be a valuable rhetorical resource, though it also lost some of its luster. Glancing back at the graphs of its rise in public discussion, we can see a fall as it became the subject of increased philosophical criticism. In 1975, Berkeley philosopher Paul Feyerabend (1924–1994) assaulted the very notion of a singular and definable scientific method in his *Against Method,* suggesting instead that scientists did whatever worked.[22] Educators, too, began to express skepticism. The 1968 edition of *Teaching Science in Today's Secondary Schools* lamented that "thousands of young people have memorized the steps" of the scientific method as they appeared in textbooks "and chanted them back to their teachers while probably doubting intuitively their appropriateness."[23] Such scrutiny cast the scientific method as narrow and brittle, depriving it of its rhetorical utility.

At the same time, the technological products of science, which had begun to invade everyday life, promised a more effective symbol of science and a bridge between the lab and the lay world. Now, instead of new scientific fields, we find biotechnology, information technology, and nanotechnology. Appeal to new technologies available in everything from electronic devices to hair products has also become a staple of advertising. Likewise, modern intellectuals routinely make use of technological metaphors, including allusions to "systems," "platforms," "constructions," or "technologies" as general methods of working. "Technoscience" has achieved widespread popularity among sociologists of science to refer to the intertwined production of abstract knowledge and material devices.

Still, the scientific method did what keywords are supposed to do. It didn't reflect reality—it helped create it. It helped to define a vision of science that was separate from other kinds of knowledge, justified the value of that science for those left on the outside, and served as a symbol of scientific prestige. It continues to accomplish those things, just not as effectively as it did during its heyday. If we return to a simplistic view, one in which the scientific method really is a recipe for producing scientific knowledge, we lose sight of a huge swath of history and the development of a pivotal touchstone on cultural maps. We deprive ourselves of a richer perspective in favor of one both narrow and contrary to the way things actually are.

THAT A CLEAR LINE OF DEMARCATION HAS

SEPARATED SCIENCE FROM PSEUDOSCIENCE

Michael D. Gordin

> Before a hypothesis can be classified as scientific, it must link to a
> general understanding of nature and conform to a cardinal rule.
> The rule is that the hypothesis must be testable. It is more
> important that there be a means of proving it *wrong* than that
> there be a means of proving it correct. On first consideration this
> may seem strange, for usually we concern ourselves with
> verifying that something is true. Scientific hypotheses are
> different. In fact, if you want to determine whether a hypothesis
> is scientific or not, look to see if there is a test for proving it
> wrong. If there is no test for its possible wrongness, then it is not
> scientific.
>
> —Paul Hewitt, *Conceptual Physics* (2002)

Quite recently, a new myth has begun to appear in science textbooks. Almost all lower-level textbooks in general science include
a section detailing "the scientific method" (see Myth 26), but now
you also find explicit discussions of what philosophers have called
"the demarcation problem": how to distinguish science from
pseudoscience. Textbooks such as Paul Hewitt's *Conceptual
Physics* consider the problem to have an obvious solution—in
order for a theory to be considered scientific, we apply a bright-
line test of "falsifiability." Whereas in earlier generations the topic
seems to have remained implicit, today falsifiability has crowded

out all possible contenders for demarcation and is considered an essential lesson for students.

Teaching students how to distinguish "real science" from impostors can reasonably be understood as *the* central task of science pedagogy. Every student in public and private schools takes several years of science, but only a small fraction of them pursue careers in the sciences. We teach the rest of them so much science so that they will appreciate what it *means* to be scientific—and, hopefully, become scientifically literate and apply some of those lessons in their lives.[1] For such students, the myth of a bright line of demarcation is essential.

The "demarcation problem" received its name in interwar Europe from philosopher Karl Popper (1902–1994), who plays an outsized role in the account that follows, but it has a venerable history—or histories. There is not one demarcation problem but several: how to distinguish correct from incorrect knowledge; how to differentiate science from all those domains (art history, theology, gardening) that are "nonscience"; and how to set science apart from things that look an awful lot like science but for some reason don't quite fit. It is this last set of supposed impostors, conventionally designated by their opponents as "pseudoscience," that are the target of the educational myth, which really only emerged explicitly in the United States since the 1980s. Both the timing and the invocation of "falsifiability" stem from the intersection of the philosophy of science with the legal debates over the teaching of creationism in the public schools.

The question of demarcation has been a central preoccupation since the earliest days of science. For example, in the fifth century BCE Hippocratic text "On the Sacred Disease," the author attacks "the sort of people we now call witch-doctors, faith-healers, quacks and charlatans," who saddled this moniker on the perfectly explicable and regular disease that moderns will come to call epilepsy.[2] Since then, the attempts by philosophers— many of whom engaged in activities we would unhesitatingly consider "science" today—to cordon off science from cuckoo's eggs have been legion and often quite ingenious.[3] They were all failures.

The problem of separating science from pseudoscience is devilishly difficult. One essential characteristic of all those doctrines, labeled as "pseudosciences," is that they very much resemble sciences, and so superficial characteristics fail to identify them.[4] We also cannot define "pseudoscience" as incorrect doctrines, because many theories that we now consider wrong—ether physics, arguments from design—were at one point unquestionably part of science (see Myth 4), which implies that many of the things we now consider to be correct science will eventually be discarded as incorrect. Are advocates of those ideas today pseudoscientific? It seems absurd to say so. The movement back and forth across the border is surprisingly vigorous, and the history of science is littered with fascinating cases (phrenology, mesmerism, acupuncture, parapsychology, and so on).[5]

As early as 1919, young Popper "*wished to distinguish between science and pseudo-science;* knowing very well that science often errs, and that pseudo-science may happen to stumble on the truth."[6] He found earlier attempts largely unsatisfactory, mostly because they considered true science to be knowledge claims that were confirmed by empirical evidence. This would never do. Three self-proclaimed "scientific" theories popular among Viennese intellectuals—the historical materialism of Karl Marx (1818–1883), the psychoanalysis of Sigmund Freud (1856–1939), and the individual psychology of Alfred Adler (1870–1937)—never lacked for confirming instances; in fact, it seems that every case presented to them could be interpreted as confirmation. Popper wondered what it would take to invalidate one of these doctrines, and was struck by the experimental confirmation of the general theory of relativity of Albert Einstein (1879–1955). During a 1919 eclipse expedition, Arthur Eddington (1882–1944) measured the bending of starlight by the gravitational field of the sun, thereby confirming the theory. Einstein had declared in 1915, when he published the theory, that if light did not exhibit the correct amount of curvature, then his theory would be incorrect. "Now the impressive thing about this case," Popper observed, "is the risk involved in a prediction of this kind." To have a chance at being right, one must gamble at being wrong—*that* was what it meant to be scientific. As he concluded: "One can sum up all this by saying that *the criterion of the scientific status of a theory is its falsifiability, or refutability, or testability.*"[7]

That is how falsifiability is usually presented, but this description truncates most of Popper's reasoning. Popper claimed that he developed these ideas in 1919 right after hearing news of the eclipse expedition, but he only coined the term "problem of demarcation" for it in 1928 or 1929, and he first unveiled the full theory at a lecture in 1953, sponsored by the British Council, on contemporary British philosophy, which he delivered at Peterhouse at Cambridge University. (Popper had left Vienna in 1937 for New Zealand and eventually the United Kingdom to escape the rise of National Socialism).[8] This history matters for two reasons: the delay further solidified the exaltation of Einstein and the denigration of Freud, making Popper sound prescient; and it was delivered in English, a few years before Popper's most significant contribution to epistemology, *Logik der Forschung* (1934), was translated as *The Logic of Scientific Discovery* (1957). In the context of Popper's full theory, falsifiability has a number of features that are rather unattractive to science educators.

For starters, Popper did not believe in truth. The bulk of his essay on falsifiability consists of a critique of the famous philosophy of induction proposed by David Hume (1711–1776). For Popper, there are no "natural laws" and nothing like "truth" in science. Instead, we have a collection of statements that have not yet been proven false. While the boldness of this position is part of its appeal, its radical skepticism is not and thus has been stripped out of typical presentations of the theory.

But even the reduced presentation of Popper's falsificationism raises serious concerns: namely, it doesn't work. Recall that Popper had (justified) concerns about how we would ever know that a theory had been confirmed; regrettably, adding a minus sign to a confirming instance does not make epistemological determination any easier. If a negative result sufficed to falsify a theory, then high school students in lab classes would have falsified pretty much everything we believe we know about the natural world. In addition, the minimum we expect of a demarcation criterion is that it group those activities we generally consider sciences in one camp, and set those commonly considered pseudosciences in another. Popper fails here, precisely because science is such a heterogeneous activity, with various methods and practices. For example, the "historical" natural sciences, such as evolutionary biology and geology—where we cannot "run the tape again"—fare poorly under the falsification test.

The situation with inclusion is even worse, as stated most forcefully in a 1983 article by the philosopher of science Larry Laudan (b. 1941):

> [Popper's criterion] has the untoward consequence of countenancing as "scientific" every crank claim that makes ascertainably false assertions. Thus flat Earthers, biblical creationists, proponents of laetrile or orgone boxes, Uri Geller devotees, Bermuda Triangulators, circle squarers, Lysenkoists, charioteers of the gods, *perpetuum mobile* builders, Big Foot searchers, Loch Nessians, faith healers, polywater dabblers, Rosicrucians, the-world-is-about-to-enders, primal screamers, water diviners, magicians, and astrologers all turn out to be scientific on Popper's criterion—just so long as they are prepared to indicate some observation, however improbable, which (if it came to pass) would cause them to change their minds.[9]

Laudan argued that *any* bright-line semantic criterion à la falsifiability would necessarily fail: demarcation was not a soluble problem. This position has been subjected to furious philosophical counterattack, yet even his critics no longer seek bright lines. Rather, they produce checklists of criteria that render a theory scientific—analogous to the *Diagnostic and Statistical Manual (DSM)*, ubiquitous in psychiatry—or groupings of "family resemblances" (following Popper's nemesis Ludwig Wittgenstein

[1889–1951]) among pseudoscientific doctrines.[10] It is almost impossible to find a philosopher of science today who thinks that Popper's criterion is the ultimate solution of the demarcation problem.

Then why do we persistently encounter this myth? The answer has less to do with philosophy or with science than with the law. Starting in the 1960s, a series of state governments in the United States passed statutes mandating "equal time" in biology courses for "evolution science" (neo-Darwinian natural selection) and "creation science" (an updated flood geology offering a scientific account that accorded closely with the creation story described in *Genesis*). Opponents countered that these laws introduced religion into the public schools, violating the constitutionally mandated separation of church and state. As one case from Arkansas reached the federal courts, the testimony of many scientists as well as philosophers and historians of science was solicited to determine the validity of the defense that creation science was a legitimate scientific hypothesis and therefore not "religion." Philosopher of science Michael Ruse (b. 1940) testified about several different demarcation criteria that would exclude scientific creationism, but one in particular impressed Judge William Overton (1939–1987) in his January 5, 1982, decision in *McLean v. Arkansas Board of Education*. In his five-point list of what makes a doctrine a "science," the final one reads: "(5) It is falsifiable (Ruse and other science witnesses)."[11] Thus, Ruse's brief sketch of Popper came to serve as a legal metric to determine whether something is scientific.

Although many philosophers of science were happy with the outcome of *McLean*, Ruse's arguments were extensively criticized, not least in Laudan's article cited earlier. Some of those critiques have stuck, and when an updated version of creationism—known as intelligent design (ID)—reached the Pennsylvania courts in 2005, Judge John Jones's (b. 1955) decision included an extensive discussion of what constituted "science," but mentioned falsifiability only twice: once in a paraphrase of Overton's decision, and once in describing how biochemist Michael Behe (b. 1952) redefined the blood-clotting mechanism to evade peer review. Instead of endorsing Popper, legal precedent now enshrines peer-reviewed publications in mainstream journals as the gold standard for demarcation.[12] We have moved from epistemology to sociology.

We have no sharp criterion for a simple reason: mimesis. Any time a test is proposed, the fringe advocates will strive to meet it *precisely* because they believe that they are pursuing proper science and agree about the need for demarcation. Creationists make plenty of falsifiable statements, for example, and now they have peer-reviewed journals. We end up with a symmetric race between the demarcators and those they wish to exclude.[13] And since the demarcation criteria have changed over time, those people accused by establishment scientists as being "pseudoscientists" bear little else in common other than their shared demonization.

Yet demarcation remains essential for the enormously high political stakes associated with climate-change denial and other antiregulatory fringe doctrines.[14] As sociologist Thomas Gieryn (b. 1950) has noted, although demarcation is a frustrating task for philosophers, for scientists it is an everyday matter: not to read this article, to ignore that email, to dismiss a website. They demarcate through socially trained judgment.[15] They do not need the myth; it's for the rest of us, who graduate from high school science classes to the ranks of registered voters.

NOTES

INTRODUCTION

Epigraph: Charles Darwin, *On the Origin of Species by Means of Natural Selection; or, The Preservation of Favoured Races in the Struggle for Life,* 6th ed. (London: John Murray, 1872), 421, with appreciation to Chip Burkhardt for bringing this to our attention.

    1. Ronald L. Numbers, ed., *Galileo Goes to Jail and Other Myths about Science and Religion* (Cambridge, MA: Harvard University Press, 2012).

    2. See, for example, John Waller's *Fabulous Science: Fact and Fiction in the History of Scientific Discovery* (Oxford: Oxford University Press, 2002), and two books by the historian of science Alberto A. Martinez: *Science Secrets: The Truth about Darwin's Finches, Einstein's Wife, and Other Myths* (Pittsburgh: University of Pittsburgh Press, 2011), and *The Cult of Pythagoras: Math and Myths* (Pittsburgh: University of Pittsburgh Press, 2012).

MYTH 1. THAT THERE WAS NO SCIENTIFIC ACTIVITY BETWEEN GREEK ANTIQUITY AND THE SCIENTIFIC REVOLUTION

Epigraph: Richard Carrier, "Christianity Was Not Responsible for Modern Science," in *The Christian Delusion,* ed. John W. Loftus (Amherst, NY: Prometheus, 2009), 414.

    1. Jim Walker, "About That Damned Graph," NoBeliefs.com, accessed April 29, 2014, http://nobeliefs.com/comments17.htm. Walker notes that for seven years he had been challenging "people to make a better graph," but that he had received no suggestions.

2. Claudio Maccone, *Mathematical SETI: Statistics, Signal Processing, and Space Missions* (Berlin: Springer, 2012), 187–188. For Sagan's time-line, see his *Cosmos* (New York: Random House, 1980), 335; I thank Neil Armstrong and apologize to Sagan for having misquoted him from memory, substituting "mankind" for "the human species," in Ronald L. Numbers, ed., *Galileo Goes to Jail and Other Myths about Science and Religion* (Cambridge, MA: Harvard University Press, 2009), 20; and David C. Lindberg and Michael H. Shank, eds., *The Cambridge History of Science*, vol. 2: *Medieval Science* (Cambridge: Cambridge University Press, 2013), 9–10.

3. We are once again in the filiation of Andrew Dickson White, *History of the Warfare of Science with Theology in Christendom,* 2 vols. (New York: D. Appleton, 1896), for whom the purported dominance of "dogmatic theology" explains the alleged absence of science in the Middle Ages.

4. See the Book Review Forum of Stephen Greenblatt, *The Swerve: How the World Became Modern* (New York: W. W. Norton, 2011), *Exemplaria* 25, no. 4 (Winter 2013): 313–370; and Hank Campbell, "*Cosmos: A Spacetime Odyssey*—the Review," Science 2.0, March 7, 2014, www.science20.com/science_20/blog/cosmos_spacetime_odyssey _review-131240.

5. See also Myths 1 and 2 in Numbers, ed., *Galileo Goes to Jail,* 8–27.

6. Americans may soon find out that direct opposition is unnecessary for science to slow down: it only takes cuts in research funds, a discourse of utility, benign neglect, and a preference for sports.

7. See the critique of Charles Homer Haskins, the distinguished historian of twelfth-century science, in Bruce Eastwood, *Ordering the Heavens: Roman Astronomy and Cosmology in the Carolingian Renaissance* (Leiden: Brill, 2007), 23–24; and H. Floris Cohen, *How Modern Science Came into the World: Four Civilizations, One 17th-Century Breakthrough* (Amsterdam: Amsterdam University Press, 2010), ch. 3.

8. Denis Feeney, *Caesar's Calendar: Ancient Time and the Beginnings of History* (Berkeley and Los Angeles: University of California Press, 2007), 196–197.

9. Of the more than one thousand papyrus rolls that Vesuvius carbonized in the Villa dei Papiri in 79, the vast majority—so far—are Greek, probably the library of the Greek philosopher Philodemus; David Sider, *The Library of the Villa dei Papiri at Herculaneum* (Los Angeles: J. Paul Getty Museum, 2003), 3–4, 43, 94–95.

10. The case of medicine is instructive; see Heinrich von Staden, "Liminal Perils: Early Modern Receptions of Greek Medicine," in *Tradition,*

Transmission, Transformation: Proceedings of Two Conferences on Pre-Modern Science Held at the University of Oklahoma, ed. F. Jamil Ragep and Sally Ragep, with Steven Livesey (Leiden: Brill, 1996): 369–418 passim, esp. 372n7, 408–409.

11. Examples include Martianus Capella, *The Marriage of Philology and Mercury,* Macrobius's *Commentary on the Dream of Scipio,* and Chalcidius's partial translation of, and commentary on, Plato's *Timaeus;* see Eastwood, *Ordering the Heavens,* chs. 2, 4–5.

12. A. I. Sabra, "The Appropriation and Subsequent Naturalization of Greek Science in Medieval Islam: A Preliminary Statement," *History of Science* 25 (1987): 223–243; Dimitri Gutas, *Greek Thought, Arabic Culture: The Graeco-Arabic Translation Movement in Baghdad and Early ʾAbbāsid Society (2nd–4th / 8th–10th Centuries)* (London: Routledge, 1998).

13. Roshdi Rashed, *Classical Mathematics from al-Khwarizmi to Descartes,* trans. M. H. Shank (London: Routledge, 2015), ch. 2.

14. See Charles Burnett, "Translation and Transmission of Greek and Islamic Science to Latin Christendom," in *The Cambridge History of Science,* 2:341–364, esp. 343, 349, 358; and Edward Grant, *A Sourcebook in Medieval Science* (Cambridge, MA: Harvard University Press, 1974), 35–38.

15. Edward Grant, "Science and the Medieval University," in *Rebirth, Reform, and Resilience: Universities in Transition, 1300–1700,* ed. James M. Kittelson and Pamela J. Transue (Columbus: Ohio State University Press, 1984), 68–102, esp. 91.

16. In contrast, the typical American undergraduate can now earn a bachelor's degree with very few introductory science courses.

17. Rainer Schwinges, *Deutsche Universitätsbesucher im 14. und 15. Jahrhundert: Studien zur Sozialgeschichte des alten Reiches* (Stuttgart: Steiner Verlag, 1986), 467–468.

18. The Museum at Alexandria is perhaps the most significant exception to this generalization.

MYTH 2. THAT BEFORE COLUMBUS, GEOGRAPHERS AND OTHER EDUCATED PEOPLE THOUGHT THE EARTH WAS FLAT

Epigraphs: Ethan Siegel, "Who Discovered the Earth Is Round?" Science Blogs, September 21, 2011, scienceblogs.com/startswithabang/2011/09/21/who-discovered-the-earth-is-ro; Gary DeMar, "Why John Kerry's Flat Earth Society Slam Is All Wrong," American Vision, May 28, 2014, americanvision.org/10905/john-kerrys-flat-earth-society-slam-wrong.

1. For earlier discussions of this myth, see Lesley B. Cormack, "Flat Earth or Round Sphere: Misconceptions of the Shape of the Earth and the Fifteenth-Century Transformation of the World," *Ecumene* 1 (1994): 363–385; and Lesley B. Cormack, "Myth 3: That Medieval Christians Taught That the Earth Was Flat," in *Galileo Goes to Jail and Other Myths about Science and Religion,* ed. Ronald L. Numbers (Cambridge, MA: Harvard University Press, 2009), 28–34, on which this essay draws.

2. Christine Garwood, *Flat Earth: The History of an Infamous Idea* (London: Macmillan, 2007), discusses some of this controversy, focusing on the "flat-earthers" of the nineteenth century.

3. William Whewell, *History of the Inductive Sciences: From the Earliest to the Present Time* (New York: D. Appleton, 1890), ch. 1, 196–197; John W. Draper, *History of the Conflict between Religion and Science* (New York: D. Appleton, 1874), 157–159. Sadly, this continues to be repeated by some textbook writers to this day. See, for example, Mounir A. Farah and Andrea Berens Karls, *World History: The Human Experience* (Lake Forest, IL: Glencoe/McGraw-Hill, 1999), and Charles R. Coble and Others, *Earth Science* (Englewood Cliffs, NJ: Prentice Hall, 1992), both intended for secondary-school audiences.

4. Washington Irving, *The Life and Voyages of Christopher Columbus: Together with the Voyages of His Companions* (London: John Murray, 1828), esp. 88.

5. Jeffrey Burton Russell, *Inventing the Flat Earth: Columbus and Modern Historians* (New York: Praeger, 1991), 24; Boies Penrose, *Travel and Discovery in the Renaissance, 1420–1620* (Cambridge, MA: Harvard University Press, 1952), 7.

6. See Charles W. Jones, "The Flat Earth," *Thought* 9 (1934): 296–307, which discusses Augustine, Jerome, Ambrose, and Lactantius.

7. Thomas Aquinas, *Summa Theologica,* par. I, qu. 47, art. 3, 1.3 (available online at www.intratext.com/IXT/ENG0023/_P1B.htm or www.Gutenberg.org). Albertus Magnus, Liber cosmographicus de natura locorum (1260), discussed in Jean Paul Tilmann, *An Appraisal of the Geographical Works of Albertus Magnus and His Contributions to Geographical Thought* (Ann Arbor: Michigan Geographical Publications, 1971). For Michael Scot, see John K. Wright, *Geographical Lore of the Time of the Crusades: A Study in the History of Medieval Science and Tradition in Western Europe* (New York: American Geographical Society, 1925), 151.

8. Walter Oakeshott, "Some Classical and Medieval Ideas in Renaissance Cosmography," in *Fritz Saxl, 1890–1948: A Volume of Memorial Essays from His Friends in England,* ed. D. J. Gordon (London: Thomas

Nelson, 1957), 251. For d'Ailly, see Arthur Percival Newton, ed., *Travel and Travellers in the Middle Ages* (London: Routledge and Kegan Paul, 1949), 14.

9. Isidore of Seville, *De Natura Rerum* 10, *Etymologiae* III 47.

10. Wesley M. Stevens, "The Figure of the Earth in Isidore's 'De Natura Rerum,'" *Isis* 71 (1980): 273. Charles W. Jones, *Bedae Opera de Temporibus* (Cambridge, MA: Medieval Academy of America, 1943), 367. See also David Woodward, "Medieval *Mappaemundi,*" in *The History of Cartography,* ed. J. B. Harley and David Woodward, vol. 1, *Cartography in Prehistoric, Ancient, and Medieval Europe and the Mediterranean* (Chicago: University of Chicago Press, 1987), 320–321.

11. Jean de Mandeville, *Mandeville's Travels,* trans. Malcolm Letts, 2 vols. (London: Hakluyt Society, 1953), 1:129.

12. Dante Alighieri, *Paradiso,* canto 9, line 84; *Inferno,* canto 26, in *The Divine Comedy,* trans. John Ciardi (New York: New American Library, 2003); Geoffrey Chaucer, *The Canterbury Tales,* in *The Works of Geoffrey Chaucer,* ed. F. N. Robinson (Boston: Houghton Mifflin, 1961), 140, line 1228.

13. Efthymios Nicolaidis, *Science and Orthodoxy: From the Greek Fathers to the Age of Globalization* (Baltimore: Johns Hopkins University Press, 2011), 24–33.

14. Most surveys of medieval science do not mention geography. David C. Lindberg, *The Beginnings of Western Science* (Chicago: University of Chicago Press, 1992), 58, devotes one paragraph to a spherical earth. J. L. E. Dreyer, *History of the Planetary Systems* (Cambridge: Cambridge University Press, 1906), 214–219, stresses Cosmas's importance, as do John H. Randall Jr., *The Making of the Modern Mind: A Survey of the Intellectual Background of the Present Age* (Boston: Houghton Mifflin, 1926), 23, and Penrose, *Travel,* who adds the caveat that "it is only fair to state that not all writers of the Dark Ages were as blind as Cosmas" (7). Jones, "Flat Earth," 305, demonstrates the marginality of Cosmas.

15. Fernando Colon, *The Life of the Admiral Christopher Columbus by His Son Ferdinand,* trans. and annotated by Benjamin Keen (Westport, CT: Greenwood Press, 1959), 39; Bartolemé de las Casas, *History of the Indies,* trans. and ed. Andrée Collard (New York: Harper and Row, 1971), 27–28.

16. Richard Eden, *The Decades of the Newe Worlde or West India . . . Wrytten in Latine Tounge by Peter Martyr of Angleria* (London, 1555), 64.

17. Concerning the long voyage, see entry for October 10, 1492, in *The Diario of Christopher Columbus's First Voyage to America, 1492–93,*

abstracted by Fray Bartolomé de Las Casas, transcribed and translated by Oliver Dunn and James E. Kelley Jr. (Norman: University of Oklahoma Press, 1989), 57. Concerning the prevailing wind, see Eden, *Decades of the Newe Worlde,* 66.

MYTH 3. THAT THE COPERNICAN REVOLUTION DEMOTED THE STATUS OF THE EARTH

I am grateful for the helpful comments of Gregory Macklem, Maurice Finocchiaro, Ronald Numbers, and Kostas Kampourakis.

Epigraph: Stephen Jay Gould, "Darwin's More Stately Mansion," *Science* 284 (1999): 2087.

1. Dennis Danielson, *The Book of the Cosmos* (New York: Basic Books, 2001), 106. This collection of sources includes excerpts of Copernicus's *On the Revolutions* (cited here) and works by other theorists cited below.

2. C. S. Lewis, *The Discarded Image* (Cambridge: Cambridge University Press, 1964), 58.

3. Danielson, *Book of the Cosmos,* 150.

4. Ibid., 171.

5. Ibid., 117, 171.

6. Johannes Kepler, *Epitome of Copernican Astronomy,* trans. Charles Wallis, vol. 16, *Great Books of the Western World* (Chicago: Encyclopedia Britannica, 1952), 848.

7. Michael J. Crowe, *The Extraterrestrial Life Debate, Antiquity to 1915: A Source Book* (Notre Dame, IN: University of Notre Dame Press, 2008), 14–34.

8. Douglas Vakoch, *Astrobiology, History, and Society* (Berlin: Springer, 2013), 341. See especially the essays in this anthology by Ted Peters and Michael Crowe.

9. Dennis Danielson, "The Great Copernican Cliché," *American Journal of Physics* 69 (2001): 1029–1035.

10. Bernard le Bovier de Fontenelle, *The Theory or System of Several New Inhabited Worlds,* trans. Aphra Behn (London: Briscoe, 1700), 16.

11. Danielson, "Great Copernican Cliché," 1033.

12. Horatio N. Robinson, *A Treatise on Astronomy* (Albany: Pease, 1849), 103.

13. Dennis Danielson and Christopher Graney, "The Case against Copernicus," *Scientific American* 310 (2013): 72–77.

14. "Bellarmine's Letter to Foscarini," in Maurice Finocchiaro, *The Essential Galileo* (Indianapolis: Hackett, 2008), 147.

15. Cecilia Payne-Gaposchkin, *Introduction to Astronomy* (New York: Prentice-Hall, 1954), 2.

16. Hermann Bondi, *Cosmology* (Cambridge: Cambridge University Press, 1952), 13.

17. Neil deGrasse Tyson, foreword to *The Cosmic Perspective*, 7th ed., by Jeffrey O. Bennett et al. (Boston: Pearson, 2014), xxviii.

18. Elaine Howard Ecklund, *Science vs. Religion: What Scientists Really Think* (New York: Oxford University Press, 2010), 57–58.

19. In a forthcoming publication I survey 130 English-language astronomy textbooks since 1621. Of those published since 2011, 78 percent contain the Copernican demotion myth. A third of those myth-perpetuating textbooks also promote a naturalistic spirituality resembling Tyson's.

20. Danielson, "Great Copernican Cliché," 1033–1034.

21. Eric Chaisson and Steve McMillan, *Astronomy: A Beginner's Guide to the Universe*, 7th ed. (Boston: Pearson, 2012), 25.

22. Guillermo Gonzalez and Donald Brownlee, "The Galactic Habitable Zone: Galactic Chemical Evolution," *Icarus* 152 (2001): 185–200; David Waltham, *Lucky Planet* (London: Icon, 2014); John Gribbin, *Alone in the Universe* (Hoboken: Wiley, 2011); Peter Ward and Donald Brownlee, *Rare Earth* (New York: Copernicus, 2000).

23. JoAnn Palmeri, "An Astronomer beyond the Observatory: Harlow Shapley as Prophet of Science" (PhD diss., University of Oklahoma, 2000), 72.

24. Mark Lupisella, "Cosmocultural Evolution: The Coevolution of Culture and Cosmos and the Creation of Cosmic Value," in *Cosmos and Culture: Cultural Evolution in a Cosmic Context,* ed. Steven Dick and Mark Lupisella (Washington, DC: NASA, 2009).

25. Tyson, foreword to *The Cosmic Perspective*, xxviii.

26. Eric Chaisson and Steve McMillan, *Astronomy Today,* 8th ed. (Boston: Pearson, 2014), 43.

MYTH 4. THAT ALCHEMY AND ASTROLOGY WERE SUPERSTITIOUS PURSUITS THAT DID NOT CONTRIBUTE TO SCIENCE AND SCIENTIFIC UNDERSTANDING

Epigraphs: George Sarton, *A History of Science,* 2 vols. (Cambridge, MA: Harvard University Press, 1952), 1:421; George Sarton, "Boyle and Bayle: The Sceptical Chymist and the Sceptical Historian," *Chymia* 3 (1950): 155–189, on 160.

1. For a fuller explanation and contextualization of alchemy and astrology in early modern culture, see Lawrence M. Principe, *The Scientific Revolution: A Very Short Introduction* (Oxford: Oxford University Press, 2011).

2. For a survey of early modern astrology, see H. Darrel Rutkin, "Astrology," in *The Cambridge History of Science,* vol. 3, *Early Modern Science,* ed. Katherine Park and Lorraine Daston (Cambridge: Cambridge University Press, 2006), 541–561.

3. Thomas Aquinas, *Summa Theologica,* IIae IIa, question 95, article 5.

4. Noel M. Swerdlow, "Galileo's Horoscopes," *Journal for the History of Astronomy* 35 (2004): 135–141; H. Darrel Rutkin, "Galileo, Astrologer: Astrology and Mathematical Practice in the Late-Sixteenth and Early-Seventeenth Centuries," *Galilaeana* 2 (2005): 107–143. Robert Boyle, *Tracts Containing Suspicions about Some Hidden Qualities of the Air; with an Appendix Touching Celestial Magnets, and Some Other Particular,* in Boyle, *Works,* ed. Michael Hunter and Edward B. Davis, 14 vols. (London: Pickering and Chatto, 1999–2000), 8:117–142.

5. Roger Bacon, *Opus Tertium,* in *Opera Quaedam Hactenus Inedita,* ed. J. S. Brewer (London, 1859), 40.

6. William R. Newman and Lawrence M. Principe, "Alchemy vs. Chemistry: The Etymological Origins of a Historiographical Mistake," *Early Science and Medicine* 3 (1998): 32–65.

7. For a full history of alchemy, see Lawrence M. Principe, *The Secrets of Alchemy* (Chicago: University of Chicago Press, 2013).

8. Ibid., 108–127.

9. Bruce Moran, *Distilling Knowledge: Alchemy, Chemistry, and the Scientific Revolution* (Cambridge, MA: Harvard University Press, 2005); Tara Nummedal, *Alchemy and Authority in the Holy Roman Empire* (Chicago: University of Chicago Press, 2007).

10. William R. Newman, *Atoms and Alchemy* (Chicago: University of Chicago Press, 2006); William R. Newman, *Promethean Ambitions: Alchemy and the Quest to Perfect Nature* (Chicago: University of Chicago Press, 2004); William R. Newman, "Technology and the Alchemical Debate in the Late Middle Ages," *Isis* 80 (1989): 423–445.

11. Lawrence M. Principe, *The Aspiring Adept: Robert Boyle and His Alchemical Quest* (Princeton, NJ: Princeton University Press, 1998); William R. Newman and Lawrence M. Principe, *Alchemy Tried in the Fire: Starkey, Boyle, and the Fate of Helmontian Chymistry* (Chicago: University of Chicago Press, 2002); Betty Jo Teeter Dobbs, *The Foundations of Newton's Alchemy* (Cambridge: Cambridge University Press, 1975). For a useful review of how current scholars see alchemy's posi-

tion in the history of science, see "Focus: Alchemy and the History of Science," *Isis* 102 (2011): 300–337.

12. Principe, *Secrets,* 83–106; Lawrence M. Principe and William R. Newman, "Some Problems in the Historiography of Alchemy," in *Secrets of Nature: Astrology and Alchemy in Early Modern Europe,* ed. Anthony Grafton and Newman (Cambridge, MA: MIT Press, 2001), 385–434.

MYTH 5. THAT GALILEO PUBLICLY REFUTED ARISTOTLE'S CONCLUSIONS ABOUT MOTION BY REPEATED EXPERIMENTS MADE FROM THE CAMPANILE OF PISA

Epigraph: Vincenzio Viviani, *Vita di Galileo* [1654], ed. Bruno Basile (Rome: Salerno, 2001), 37–38.

1. Vincenzio Viviani, *Vita di Galileo* [1654], ed. Bruno Basile (Rome: Salerno, 2001), 37–38.

2. Lane Cooper, *Aristotle, Galileo, and the Tower of Pisa* (Ithaca, NY: Cornell University Press, 1935), 13–26, gives several remarkable examples. Cooper, a professor of English at Cornell, was perhaps the first to expose Viviani's story as a fable.

3. J. L. Heilbron, *Galileo* (Oxford: Oxford University Press, 2010), 41–45; R. V. Caffarelli in Caffarelli, ed., *Galileo e Pisa* (Pisa: Felice, 2004), 40, citing Thomas Settle for the filmed confirmation.

4. Galileo Galilei, *Opere,* ed. Antonio Favaro, 20 vols. (Florence: Giunti-Barbera, 1890–1910), 1:334.

5. Galileo, re Giorgio Coresio, professor of Greek at Pisa, in *Opere,* 4:285.

6. Aristotle, *Physica,* 4.1–4.5, esp. 210a30–211a1–5, 212a5–10, 20, in *The Works of Aristotle,* ed. David Ross, trans. R. P. Hardie and R. K. Gaye, 12 vols. (Oxford: Oxford University Press, 1928–1952), vol. 3.

7. Aristotle, *Physica,* 4.8, 214b30–35, 215a19–21.

8. Ibid., 214a25–30, 216a8–11.

9. Ibid., 216a8–21.

10. Galileo, *Opere,* 8:108–109.

11. Heilbron, *Galileo,* 50–51 (quotations).

12. Renieri to Galileo, 13 and 20 Mar 1641, in *Opere,* 18:305–306, 310.

13. For Mazzoni, see Heilbron, *Galileo,* 14–15, 46–8, 53–54, 110–112.

14. Examples are given throughout Heilbron, *Galileo,* esp. 254–252, 270–276.

MYTH 6. THAT THE APPLE FELL AND NEWTON INVENTED THE LAW OF GRAVITY, THUS REMOVING GOD FROM THE COSMOS

I am grateful to the editors for inviting me to contribute to this book and for their helpful comments on draft versions. I should also like to thank Steven Snobelen for his valuable advice.

Epigraphs: Steven Weinberg, "On God, Christianity and Islam," *Times Literary Supplement* (17 January 2007); Johnjoe McFadden, "Survival of the Wisest," *Guardian,* 30 June 2008.

    1. Patricia Fara, *Newton: The Making of Genius* (London: Macmillan, 2002), 192.

    2. William Stukeley, *Memoirs of Sir Isaac Newton's Life* (London: Taylor and Francis, 1936; originally published in 1752), 20.

    3. Isaac Newton, *Philosophiæ Naturalis Principia Mathematica* (London, 1687).

    4. Jean-Baptiste Biot, "Newton," trans. Howard Elphinstone, in *Lives of Eminent Persons* (London: Baldwin and Cradock, 1833).

    5. Isaac Newton, *Mathematical Principles of Natural Philosophy,* trans. Andrew Motte, 2 vols. (London, 1729), 2:390–391.

    6. Ibid., 2:391–392. This is the 1726 version; in 1713, he wrote "Experimental Philosophy."

    7. Joseph Warton et al., eds., *The Works of Alexander Pope,* 9 vols. (London: Richard Priestley, 1822), 2:379.

MYTH 7. THAT FRIEDRICH WÖHLER'S SYNTHESIS OF UREA IN 1828 DESTROYED VITALISM AND GAVE RISE TO ORGANIC CHEMISTRY

Epigraphs: Wikipedia, s.v. "Wöhler Synthesis," accessed August 28, 2014, http://en.wikipedia.org/wiki/Wöhler_synthesis; David Klein, *Organic Chemistry* (New York: Wiley, 2012), 2.

    1. Friedrich Wöhler, "Ueber Künstliche Bildung des Harnstoffs," *Annalen der Physik und Chemie* 88 (1828): 253–256.

    2. Peter J. Ramberg, "The Death of Vitalism and the Birth of Organic Chemistry: Wöhler's Urea Synthesis in Textbooks of Organic Chemistry," *Ambix* 47 (2000): 170–195.

    3. Douglas McKie, "Wöhler's Synthetic Urea and the Rejection of Vitalism: A Chemical Legend," *Nature* 153 (1944): 608–610, on 608.

    4. J. H. Brooke, "Wöhler's Urea, and Its Vital Force: A Verdict from the Chemists," *Ambix* 15 (1968): 84–114.

5. Michel Eugène Chevreul, *A Chemical Study of Oils and Fats of Animal Origin*, trans. Gary R. List and Jaime Wisniak (Urbana, IL: AOCS Press, 2009).

6. Albert B. Costa, *Michel Eugène Chevreul, Pioneer of Organic Chemistry* (Madison: Wisconsin State Historical Society, 1962).

7. Jöns Jakob Berzelius, "Experiments to Determine the Definite Proportions in Which the Elements of Organic Nature Are Combined," *Annals of Philosophy* 4 (1814): 323–331, on 323.

8. John Hedley Brooke, "Berzelius, the Dualistic Hypothesis, and the Rise of Organic Chemistry," in *Enlightenment Science in the Romantic Era: The Chemistry of Berzelius and Its Cultural Setting*, ed. Evan Melhado and Tore Frangsmyer (Cambridge: Cambridge University Press, 1992), 180–221.

9. Ibid., 188.

10. John Hedley Brooke, "Organic Synthesis and the Unification of Chemistry: A Reappraisal," *British Journal for the History of Science* 5 (1971): 363–392.

11. Julien Offray de La Mettrie, *L'Homme Machine* (Leyden: Elie Luzac, 1748).

12. Timothy Lenoir, *The Strategy of Life: Teleology and Mechanics in Nineteenth-Century German History of Biology* (Chicago: University of Chicago Press, 1989).

13. Alan J. Rocke, "Berzelius' Animal Chemistry: From Physiology to Organic Chemistry (1805–1814)," in Melhado and Frangsmyer, *Enlightenment Science*, 107–131; Bent Søren Jørgensen, "More on Berzelius and the Vital Force," *Journal of Chemical Education* 42 (1965): 394–396.

14. Justus von Liebig, *Animal Chemistry; or, Organic Chemistry in Its Application to Physiology and Pathology* (Cambridge, MA: John Owen, 1842), 209.

15. Timothy Lipman, "Wöhler's Preparation of Urea and the Fate of Vitalism," *Journal of Chemical Education* 41 (1964): 452–458.

16. Garland Allen, "Mechanism, Vitalism and Organicism in Late Nineteenth and Twentieth-Century Biology: The Importance of Historical Context," *Studies in History and Philosophy of Biological and Biomedical Sciences* 36 (2005): 261–283.

17. Scott Gilbert and Sahotra Sarkar, "Embracing Complexity: Organicism for the 21st Century," *Developmental Dynamics* 219 (2000): 1–9.

18. Quoted in John Farley and Gerald Geison, "Science, Politics and Spontaneous Generation in Nineteenth-Century France: The

Pasteur-Pouchet Debate," *Bulletin of the History of Medicine* 48 (1974): 161–198, on 177.

19. Ibid.

20. Quoted in Paolo Palladino, "Stereochemistry and the Nature of Life: Mechanist, Vitalist, and Evolutionary Perspectives," *Isis* 81 (1990): 44–67, on 52.

21. The nationalistic origins of the myth help explain why Chevreul's study of animal and vegetable fats, or Marcelin Berthelot's (1827–1897) total syntheses—nearly all done directly from the elements—have been considered as only "paving the way" for, or substantiating, Wöhler's urea synthesis. See Ramberg, "Death of Vitalism."

22. Allen, "Mechanism, Vitalism and Organicism."

MYTH 8. THAT WILLIAM PALEY RAISED SCIENTIFIC QUESTIONS ABOUT BIOLOGICAL ORIGINS THAT WERE EVENTUALLY ANSWERED BY CHARLES DARWIN

Epigraphs: Richard Dawkins, *The Blind Watchmaker* (Harlow, Essex: Longman Scientific and Technical, 1986), 5; Michael Behe, *Darwin's Black Box* (New York: Free Press, 1996), 211.

1. On an eternal universe, see Brian Stock, "Science, Technology, and Economic Progress in the Early Middle Ages," in *Science in the Middle Ages,* ed. David Lindberg (Chicago: University of Chicago Press, 1976), 42–43; William Paley, *Natural Theology; or, Evidences of the Existence and Attributes of the Deity Collected from the Appearances of Nature* (London: R. Faulder 1802), 1.

2. Paley, *Natural Theology,* 9.

3. Ibid., 437–586; Adam R. Shapiro, "William Paley's Lost 'Intelligent Design,'" *History and Philosophy of the Life Sciences* 31 (2009): 55–78.

4. Paley, *Natural Theology,* 9. The themes of toleration and natural religion are emphasized in A Friend of Religious Liberty [William Paley], *A Defence of the Considerations on the Propriety of Requiring a Subscription to Articles of Faith, in Reply to a Late Answer from the Clarendon Press* (London: J. Wilkie, 1774); and William Paley, "Divine Benevolence," in *Principles of Moral and Political Philosophy* (London: R. Faulder, 1785).

5. Neal C. Gillespie, "Divine Design and the Industrial Revolution: William Paley's Abortive Reform of Natural Theology," *Isis* 81 (1990): 214–229; Wilson Smith, "William Paley's Theological Utilitarianism in America," *William and Mary Quarterly* 11 (1954): 402–424; Kevin Gilmartin, "In the Theater of Counterrevolution: Loyalist Association

and Conservative Opinion in the 1790s," *Journal of British Studies* 41 (2002): 291–328.

6. Paley, *Moral and Political Philosophy*. Natural evidence for moral laws is also evident in Paley, *Reasons for Contentment, Addressed to the Labouring Part of the British Public* (Carlisle: F. Jollie, 1792).

7. Henry Lord Brougham and Sir Charles Bell, *Paley's Natural Theology with Illustrative Notes* (London: Charles Knight, 1836), 1–2.

8. Paley, *Natural Theology*, 463; Paul L. Farber, "Buffon and the Concept of Species," *Journal of the History of Biology* 5 (1972): 259–284; James R. Moore, *The Post-Darwinian Controversies: A Study of the Protestant Struggle to Come to Terms with Darwin in Great Britain and America, 1870–1900* (Cambridge: Cambridge University Press, 1979), 310–311. See also Myths 10–14 in this volume.

9. Paley, *Natural Theology*, 464–465.

10. James G. Lennox, "Darwin *was* a Teleologist," *Biology and Philosophy* 8 (1993): 409–421.

11. Charles Darwin, *On the Origin of Species; or, the Preservation of Favoured Races in the Struggle for Life* (London: John Murray, 1859), 201.

12. Adam R. Shapiro, "Darwin's Foil: The Evolving Uses of William Paley's Natural Theology, 1802–2005," *Studies in History and Philosophy of Biological and Biomedical Sciences* 45 (2014): 114–123.

MYTH 9. THAT NINETEENTH-CENTURY GEOLOGISTS WERE DIVIDED INTO OPPOSING CAMPS OF CATASTROPHISTS AND UNIFORMITARIANS

I would like to acknowledge the very useful guidance and feedback of Kostas Kampourakis, Ronald L. Numbers, and Nicolaas Rupke.

Epigraphs: William Whewell, review of *Principles of Geology: Being an Attempt to Explain the Former Changes of the Earth's Surface, by Reference to Causes Now in Operation*, vol. 2, by Charles Lyell, *Quarterly Review* 47 (1832): 103–132, on 126; Clarence King, "Catastrophism and Evolution," *American Naturalist* 11 (1877): 449–470, on 451–452.

1. George W. White, introduction and biographical notes to *Illustrations of the Huttonian Theory of the Earth*, by John Playfair (1802; repr., New York: Dover Publications, 1964), v–xix, v–vi.

2. Playfair, *Illustrations*, 3.

3. Charles Lyell, *Principles of Geology: Being an Attempt to Explain the Former Changes of the Earth's Surface, by Reference to Causes Now in Operation*, vol. 1 (London: John Murray, 1830). Volume 2 appeared in 1832; volume 3 in 1833. See also Martin J. S. Rudwick, "The Strategy

of Lyell's *Principles of Geology,*" *Isis* 61 (1970): 4–33; and Martin J. S. Rudwick, "Uniformity and Progression: Reflections on the Structure of Geological Theory in the Age of Lyell," in *Perspectives in the History of Science and Technology,* ed. Duane H. D. Roller (Norman: University of Oklahoma, 1971), 209–228.

4. W[illiam] D[aniel] Conybeare, "Report on the Progress, Actual State, and Ulterior Prospects of Geological Science," *Second Report of the British Association for the Advancement of Science* (1832): 365–414, on 406. See also [Adam] Sedgwick, "Address to the Geological Society, Delivered on the Evening of the 18th of February 1831," *Proceedings of the Geological Society of London,* no. 20 (1831): 281–316.

5. Whewell, review of *Principles of Geology,* 126. For an American example, see James Dwight Dana, "On the Analogies between the Modern Igneous Rocks and the So-Called Primary Formations and the Metamorphic Changes Produced by Heat in the Associated Sedimentary Deposits," *American Journal of Science* 45 (1843): 104–129, on 104.

6. Trevor Palmer, *Perilous Planet Earth: Catastrophes and Catastrophism through the Ages* (Cambridge: Cambridge University Press, 2003), ch. 5.

7. Martin J. S. Rudwick, *Worlds before Adam: The Reconstruction of Geohistory in the Age of Reform* (Chicago: University of Chicago Press, 2008), 291–292, 294, 563–564; Rudwick, "Strategy," 9–10; M. J. S. Rudwick, "Lyell and the *Principles of Geology,*" in *Lyell: The Past Is the Key to the Present,* ed. D. J. Blundell and A. C. Scott, Special Publications, vol. 143, (London: Geological Society, 1998), 3–15, 7.

8. Michael Bartholomew, "The Singularity of Lyell," *History of Science* 17 (1979): 276–293; "Principles of Geology," *American Journal of Science* 42 (1842): 191–192.

9. Julie Renee Newell, "American Geologists and Their Geology: The Formation of the American Geological Community, 1780–1865 (PhD diss., University of Wisconsin-Madison, 1993), chs. 4, 6.

10. See, for instance, Derek Ager, *The New Catastrophism: The Importance of the Rare Event in Geological History* (Cambridge: Cambridge University Press, 1993); Nicolaas A. Rupke, "Reclaiming Science for Creationism," in *Creationism in Europe,* ed. Stefaan Blancke, Hans Henrik Hjermitslev, and Peter C. Kjærgaard (Baltimore: Johns Hopkins University Press, 2014), 242–249; and Palmer, *Perilous Planet Earth.*

MYTH 10. THAT LAMARCKIAN EVOLUTION RELIED LARGELY ON USE AND DISUSE AND THAT DARWIN REJECTED LAMARCKIAN MECHANISMS

Epigraph: Peter H. Raven and George B. Johnson, *Biology,* 6th ed. (Boston: McGraw-Hill, 2002), 422.

1. Charles Darwin, *On the Origin of Species by Means of Natural Selection,* 6th ed. (London: John Murray, 1872), 421.

2. Charles Darwin, *The Variation of Animals and Plants under Domestication,* 2 vols. (London: John Murray, 1868), 2:ch. 27.

3. Darwin, *Origin,* 176; Charles Darwin, "Sir Wyville Thomson and Natural Selection," *Nature* 23 (November 1880): 32.

4. On Lamarck, see especially Richard W. Burkhardt Jr., *The Spirit of System: Lamarck and Evolutionary Biology* (Cambridge, MA: Harvard University Press, 1977); and Pietro Corsi, *The Age of Lamarck: Evolutionary Theories in France, 1790–1830* (Berkeley and Los Angeles: University of California Press, 1988).

5. Jean-Baptiste Lamarck, *Système des Animaux sans Vertèbres* (Paris: Déterville, 1801).

6. Jean-Baptiste Lamarck, *Recherches sur l'Organisation des Corps Vivans* (Paris: Maillard, 1802), 38.

7. Jean-Baptiste Lamarck, *Philosophie Zoologique,* 2 vols. (Paris: Dentu, 1809), 1:221.

8. Jean-Baptiste Lamarck, *Histoire Naturelle des Animaux sans Vertèbres,* 7 vols. (Paris: Déterville, 1815–1822), 1:132–133.

9. Burkhardt, *Spirit of System,* 151–157.

10. Lamarck, *Système des Animaux,* 13.

11. Lamarck, *Philosophie Zoologique,* 1:235.

12. Lamarck, *Histoire Naturelle,* 1:200.

13. Richard W. Burkhardt Jr., "Lamarck, Cuvier, and Darwin on Animal Behavior and Acquired Characters," in *Transformations of Lamarckism: From Subtle Fluids to Molecular Biology,* ed. Snait B. Gissis and Eva Jablonka (Cambridge MA: MIT Press, 2011), 33–44; Richard W. Burkhardt Jr., "Lamarck, Evolution, and the Inheritance of Acquired Characters," *Genetics* 194 (2013): 793–805.

14. Lamarck, *Histoire Naturelle,* 1:191.

15. Charles Darwin, *On the Origin of Species by Means of Natural Selection* (London: John Murray, 1859), 134.

16. Darwin, *Origin* (1859), 11, 137.

17. Lamarck, *Philosophie Zoologique,* 1:241, 260.

18. Darwin, *Origin* (1859), 242.

19. Darwin, *Variation,* 2:395.

20. Darwin, *Origin,* 6th ed. (1872), 421.

21. For an instructive analysis of the multiple ways in which the term "Lamarckian" is commonly used (and misused), see Kostas Kampourakis and Vasso Zogza, "Students' Preconceptions about Evolution: How Accurate Is the Characterization as 'Lamarckian' When Considering the History of Evolutionary Thought?" *Science & Education* 16 (2007): 393–422.

MYTH 11. THAT DARWIN WORKED ON HIS THEORY IN SECRET FOR TWENTY YEARS, HIS FEARS CAUSING HIM TO DELAY PUBLICATION

Epigraphs: Michael Ruse, *Defining Darwin* (Amherst, NY: Prometheus Books, 2009), 72; Howard Gruber, *Darwin on Man* (New York: Dutton, 1974), xiv.

1. Robert J. Richards, "Why Darwin Delayed, or Interesting Problems and Models in the History of Science," *Journal of the History of the Behavioral Sciences* 19 (1983): 45–53.

2. John Greene, *The Death of Adam: Evolution and Its Impact on Western Thought* (Ames: Iowa State University Press, 1959), 260.

3. J. W. Burrow, introduction to *The Origin of Species by Means of Natural Selection; or, The Preservation of Favoured Races in the Struggle for Life,* by Charles Darwin (Baltimore: Penguin Books, 1968), 32.

4. Michael Ruse, *The Darwinian Revolution: Science Red in Tooth and Claw* (Chicago: University of Chicago Press, 1979), 185.

5. Gruber, *Darwin on Man,* 202.

6. Stephen Jay Gould, *Ever Since Darwin* (New York: Norton, 1977), 21–27. The essay was first published in Gould's column "This View of Life," *Natural History Magazine* 83 (December 1974): 68.

7. Adrian Desmond and James Moore, *Darwin: The Life of a Tormented Evolutionist* (New York: Norton, 1994; originally published in 1991), xviii.

8. Adam Gopnik, "Rewriting Nature: Charles Darwin, Natural Novelist," *New Yorker,* October 23, 2006, 52–59.

9. The manuscript is held in the Manuscript Room, Cambridge University Library, DAR 73.

10. Charles Darwin, *On the Origin of Species* (London: Murray, 1859), 236.

11. See my *Darwin and the Emergence of Evolutionary Theories of Mind and Behavior* (Chicago: University of Chicago Press, 1987), 142–152.

12. Charles Darwin, *The Autobiography of Charles Darwin, 1809–1882*, ed. Nora Barlow (New York: Norton, 1969), 120–121; Darwin, *Origin of Species*, 280.

13. See my "Darwin's Principle of Divergence: Why Fodor Was Almost Right," in my *Was Hitler a Darwinian?* (Chicago: University of Chicago Press, 2013), 55–89.

14. Ibid., 61–65.

15. Darwin, *Origin*, 459.

16. Charles Darwin, "Observations on the Parallel Roads of Glen Roy," *Philosophical Transactions of the Royal Society of London*, pt. 1 (1839): 39–81.

17. Darwin, *Autobiography*, 84.

18. John van Wyhe, "Mind the Gap: Did Darwin Avoid Publishing His Theory for Many Years?" *Notes and Records of the Royal Society* 61 (2007): 177–205.

19. Ibid., 178.

20. Charles Darwin, *Notebook C* (MS 123 and 202), in *Charles Darwin's Notebooks, 1836–1844*, ed. Paul Barrett et al. (Ithaca, NY: Cornell University Press, 1987), 276, 302; brackets indicate Darwin's insertion.

21. Charles Darwin, *Notebook M* (MS 19 and 57), *Old and Useless Notes* (MS 37, 39, 49v), and *Notebook C* (MS 166), in *Charles Darwin's Notebooks*, 524, 532–533, 614, 616, 618, 291.

22. Charles Darwin to Joseph Hooker (11 January 1844), in *Correspondence of Charles Darwin*, 22 volumes to date (Cambridge: Cambridge University Press, 1985–), 3:2.

23. John Bowlby, *Charles Darwin: A New Life* (New York: Norton, 1990), 254–255, 323.

24. Charles Darwin and Alfred Wallace, "On the Tendency of Species to Form Varieties; and on the Perpetuation of Varieties and Species by Natural Means of Selection," *Journal of the Proceedings of the Linnean Society: Zoology* 3 (August 1858): 45–62.

MYTH 12. THAT WALLACE'S AND DARWIN'S EXPLANATIONS OF EVOLUTION WERE VIRTUALLY THE SAME

Epigraphs: George Ledyard Stebbins, *Processes of Organic Evolution* (Englewood Cliffs, NJ: Prentice-Hall, 1971), 8; C. R. Darwin to Charles Lyell, 18 June 1858, Darwin Correspondence Database, www.darwinproject.ac.uk/entry-2285.

1. Charles Darwin, *On the Origin of Species by Means of Natural Selection; or, The Preservation of Favoured Races in the Struggle for Life* (London: John Murray, 1859); Charles Darwin and Alfred Russel Wallace, *Evolution by Natural Selection*, with a foreword by Gavin de Beer (Cambridge: Cambridge University Press, 1958). The latter book contains Darwin's early expositions of his theory and the Darwin-Wallace papers of 1858 presented at the Linnean Society.

2. Peter J. Bowler, *Darwin Deleted: Imagining a World without Darwin* (Chicago: University of Chicago Press, 2013).

3. Michael Ruse, *The Darwinian Revolution: Science Red in Tooth and Claw,* 2nd ed. (Chicago: University of Chicago Press, 1999).

4. John Frederick William Herschel, *Preliminary Discourse on the Study of Natural Philosophy* (London: Longman, Rees, Orme, Brown, and Green, 1830).

5. Michael Ruse, *Darwin and Design: Does Evolution Have a Purpose?* (Cambridge, MA: Harvard University Press, 2003).

6. Peter J. Bowler, *Evolution: The History of an Idea* (Berkeley and Los Angeles: University of California Press, 1984).

7. Darwin, *Origin of Species,* 63.

8. Adam Smith, *The Glasgow Edition of the Works and Correspondence of Adam Smith,* ed. R. H. Campbell and A. S. Skinner (Oxford: Clarendon Press, 1976), 2A, 26–27.

9. Michael Ruse, "Charles Darwin and Group Selection," *Annals of Science* 37 (1980): 615–630.

10. Charles Darwin, *The Descent of Man, and Selection in Relation to Sex* (London: John Murray, 1871).

11. Alfred Russel Wallace, *Contributions to the Theory of Natural Selection* (London: Macmillan, 1870). Interestingly, Wallace later argued that there is—or will be—female choice among humans, as young women mate only with the best young men. This strikes me as even more implausible than spiritualism. See Alfred Russel Wallace, *Studies: Scientific and Social* (London: Macmillan, 1900).

MYTH 13. THAT DARWINIAN NATURAL SELECTION HAS BEEN "THE ONLY GAME IN TOWN"

Epigraphs: Jerry A. Coyne, *Why Evolution Is True* (New York: Penguin, 2009), xvi; Richard Dawkins, *The Greatest Show on Earth: The Evidence for Evolution* (New York: Free Press, 2009), 426.

1. Charles Darwin, "An Historical Sketch on the Progress of Opinion on the Origin of Species Previously to the Publication of the First Edi-

tion of This Work," *The Origin of Species*, 4th ed. (London: John Murray, 1866), xiii.

2. Bentley Glass, preface to *Forerunners of Darwin: 1745–1859*, ed. Bentley Glass, Owsei Temkin, and William L. Straus Jr. (Baltimore: Johns Hopkins University Press, 1959), vi.

3. The early twenty-first century is exhibiting a growth of interest in the history of "deviations from Darwin." See, for example, Abigail Lustig, Robert J. Richards, and Michael Ruse, eds., *Darwinian Heresies* (Cambridge: Cambridge University Press, 2004); George S. Levit, Kay Meister, and Uwe Hoßfeld, "Alternative Evolutionary Theories: A Historical Survey," *Journal of Bioeconomics* 10 (2008): 71–96; Peter J. Bowler, *Darwin Deleted: Imagining a World without Darwin* (Chicago: University of Chicago Press, 2013).

4. Nicolaas Rupke, *Richard Owen: Biology without Darwin* (Chicago: University of Chicago Press, 2009), 161–165.

5. Dawkins, *Greatest Show on Earth*, 326.

6. See, for example, Jerry Fodor and Massimo Piatelli-Palmarini, *What Darwin Got Wrong* (New York: Farrar, Straus and Giroux, 2010), xiii.

7. By contrast, Darwin narrowed the scope of evolutionary biology by belittling the problem of how life had originated. Both Ernst Mayr and Richard Dawkins have reasserted the separateness and uniqueness of biology, keeping the conundrum of life's origins at arm's length; both men have also doubted the validity of astrobiology with its implications of multiple spontaneous origins of life. On Darwin's stance, see Nicolaas Rupke, "Darwin's Choice," in *Biology and Ideology from Descartes to Darwin*, ed. Denis R. Alexander and Ronald L. Numbers (Chicago: University of Chicago Press, 2014), 139–164.

8. Simon Conway Morris, ed., *The Deep Structure of Biology: Is Convergence Sufficiently Ubiquitous to Give a Directional Signal?* (West Conshohocken, PA: Templeton Foundation Press, 2008); Keith Bennett, "The Chaos Theory of Evolution," *New Scientist* 208 (October 2010): 28–31, on 31.

9. Nicolaas Rupke, "The Origin of Species from Linnaeus to Darwin," in *Aurora Torealis*, ed. Marco Beretta, Karl Grandin, and Svante Lindquist (Sagamore Beach, MA: Science History Publications, 2008), 71–85.

10. Wentworth D'Arcy Thompson, *On Growth and Form* (Cambridge: Cambridge University Press, 1917).

11. Simon Conway Morris, *Life's Solution: Inevitable Humans in a Lonely Universe* (Cambridge: Cambridge University Press, 2003); The "lonely universe" part of Conway Morris's argument is, of course, not

structuralist but joins Darwin in a mystifying approach to abiogenesis, which was an accommodation not only to the limitations of his theory of natural selection but also to the religious sensibilities of his contemporaries.

12. David N. Livingstone, *Dealing with Darwin: Place, Politics and Rhetoric in Religious Engagements with Evolution* (Baltimore: Johns Hopkins University Press, 2014).

13. John C. Greene, *Science, Ideology and World View* (Berkeley and Los Angeles: University of California Press, 1981), 7.

14. In Scotland (Edinburgh) and Ireland (both Belfast and Dublin), German structuralism had strong scientific representation and held its ground until well after the appearance of *The Origin of Species*.

15. Lynn K. Nyhart, *Biology Takes Form: Animal Morphology and the German Universities, 1800–1900* (Chicago: University of Chicago Press, 1995).

16. Nicolaas Rupke, *Alexander von Humboldt: A Metabiography* (Chicago: University of Chicago Press, 2008), 88–92.

17. Stephen Jay Gould, foreword to *Basic Questions in Paleontology: Geologic Time, Organic Evolution, and Biological Systematics*, by Otto H. Schindewolf (Chicago: University of Chicago Press, 1993), xi.

18. Richard Dawkins, *Climbing Mount Improbable* (New York: W. W. Norton, 1996), 28.

19. Ernst Mayr, *The Growth of Biological Thought : Diversity, Evolution and Inheritance* (Cambridge MA : Harvard University Press, 1982).

20. This chapter is a condensed digest of the history of the structuralist approach to organic evolution on which I am currently working, provisionally entitled *Roll Over, Darwin: The Structuralist Tradition in Evolutionary Biology*.

MYTH 14. THAT AFTER DARWIN (1871), SEXUAL SELECTION WAS LARGELY IGNORED UNTIL ROBERT TRIVERS (1972) RESURRECTED THE THEORY

Thanks to Kostas Kampourakis and Ronald Numbers for organizing this volume and their keen editorial advice, as well as to Dick Burian for his thoughtful comments on an earlier version of this chapter. The engaging conversation at the Historical Myths about Science conference coordinated by Nicolaas Rupke enriched my thinking about sexual selection and many other topics as well; thank you to all the participants.

Epigraph: John Alcock, *Animal Behavior: An Evolutionary Approach,* 4th ed. (Sunderland, MA: Sinauer, 1989), 398.

1. Man: A Course of Study, *Animal Adaptation* (Washington, DC: Curriculum Development Associates, 1970); Man: A Course of Study, *Natural Selection* (Washington, DC: Curriculum Development Associates, 1970); Man: A Course of Study, *Innate and Learned Behavior* (Washington, DC: Curriculum Development Associates, 1970).

2. Beginning in the 1970s, botanists also explored whether sexual selection theory might provide a useful set of theoretical tools for thinking about variation in fertilization success in plants, but their research has attracted less scholarly attention; see, for example, Mary F. Willson, "Sexual Selection in Plants," *American Naturalist* 113 (1979): 777-790.

3. For Charles Darwin's most thorough explanation of the differences between natural and sexual selection, see *Descent of Man and Selection in Relation to Sex,* 2 vols. (London: John Murray, 1871).

4. Darwin first described sexual selection in *On the Origin of Species* (London: John Murray, 1859).

5. See, for example, Henrika Kuklick, *The Savage Within: The Social History of British Anthropology* (Cambridge: Cambridge University Press, 1993); and B. Ricardo Brown, *Until Darwin: Science, Human Variety and the Origins of Race* (London: Pickering and Chatto, 2010).

6. For example, Angelique Richardson, *Love and Eugenics in the Late Nineteenth Century: Rational Reproduction and the New Woman* (New York: Oxford University Press, 2008); Erika Lorraine Milam, *Looking for a Few Good Males: Female Choice in Evolutionary Biology* (Baltimore: Johns Hopkins University Press, 2010); and Kimberly Hamlin, *From Eve to Evolution: Darwin, Science, and Women's Rights in Gilded Age America* (Chicago: University of Chicago Press, 2014).

7. On laboratory studies of sexual selection in these years, see Erika Lorraine Milam, "'The Experimental Animal from the Naturalist's Point of View': Evolution and Behavior at the AMNH, 1928-1954," in *Descended from Darwin: Insights into the History of Evolutionary Studies, 1900-1970,* ed. Joe Cain and Michael Ruse, Transactions of the American Philosophical Society, vol. 99, pt. 1 (Philadelphia: American Philosophical Society, 2009), 157-178. On the scientific authority of *Drosophila,* see Robert Kohler, *Lords of the Fly:* Drosophila *Genetics and the Experimental Life* (Chicago: University of Chicago Press, 1994).

8. Milam, *Looking for a Few Good Males,* ch. 5.

9. Robert Trivers, "The Evolution of Reciprocal Altruism," *Quarterly Review of Biology* 46 (1971): 35-57; quotation from Robert Trivers,

Natural Selection and Social Theory: Selected Papers of Robert Trivers (New York: Oxford University Press, 2002), 5.

10. See, for example, Marcel Mauss, *The Gift,* trans. W. D. Halls (London: Routledge, 1990).

11. On the declining fortunes of group selection during this period, see Mark Borello, *Evolutionary Restraints: The Contentious History of Group Selection* (Chicago: University of Chicago Press, 2010).

12. Robert Trivers, "Parental Investment and Sexual Selection," in *Sexual Selection and the Descent of Man, 1871–1971,* ed. Bernard Campbell (Chicago: Aldine, 1972), 136–179.

13. Trivers's two other publications of the early 1970s were Robert Trivers and Daniel Willard, "Natural Selection of Parental Ability to Vary the Sex Ratio of Offspring," *Science* 179 (1973): 90–92; and Robert Trivers, "Parent-Offspring Conflict," *American Zoologist* 14 (1974): 247–262.

14. Alcock, *Animal Behavior: An Evolutionary Approach* (Sunderland, MA: Sinauer, 1975); 2nd edition (1979); 3rd edition (1984); 4th edition (1989); 5th edition (1993); 6th edition (1998); 7th edition (2001); 8th edition (2005); 9th edition (2009).

15. E. O. Wilson, *Sociobiology: A New Synthesis* (Cambridge, MA: Harvard University Press, 1975); Paul Erickson, "Mathematical Models, Rational Choice, and the Search for Cold War Culture," *Isis* 101 (2010): 386–392.

16. In contrast, Douglas Futuyma produced an equally popular textbook, *Evolutionary Biology,* in which he treated sexual selection as a subset of natural selection, and used the final chapter of the book to argue against the application of evolutionary theory to human behavior. Futuyma, *Evolutionary Biology* (Sunderland, MA: Sinauer, 1979).

17. Alcock, *Animal Behavior,* 2nd ed., 259.

18. Alcock, *Animal Behavior,* 3rd ed., 380.

19. On "great men" and heroic narratives in the history of science, see Misia Landau, "Human Evolution as Narrative," *American Scientist* 72 (1984): 262–268; Mary Jo Nye, "Scientific Biography: History of Science by Another Means," *Isis* 97 (2006): 322–329; and Mott Greene, "Writing Scientific Biography," *Journal of the History of Biology* 40 (2007): 727–759. If the conventional structures of scientific biographies have changed little in the past decades, neither have the means by which textbook authors associate scientific theories with the identity of the "discoverers."

20. The wording, however, remained identical; Alcock, *Animal Behavior,* 5th ed., 400, 402.

21. Alcock, *Animal Behavior,* 6th ed., 439.

22. See the self-reflective essays in Trivers, *Natural Selection and Social Theory.*

23. In the eighth and ninth editions, Alcock softened even Darwin's omniscience, writing, "The realization that the members of a species can determine who gets to reproduce and who fails to do so led Charles Darwin to propose that evolutionary chance could be driven by sexual selection." Alcock, *Animal Behavior,* 8th ed., 338; Alcock, *Animal Behavior,* 9th ed., 340.

24. See, for example, the storm of controversy engendered by Joan Roughgarden's challenge to the field: Joan Roughgarden, *The Genial Gene: Deconstructing Darwinian Selfishness* (Berkeley and Los Angeles: University of California Press, 2009).

25. Robert Trivers, *Folly of Fools: The Logic of Deceit and Self-Deception in Human Life* (New York: Basic Books, 2011).

MYTH 15. THAT LOUIS PASTEUR DISPROVED SPONTANEOUS GENERATION ON THE BASIS OF SCIENTIFIC OBJECTIVITY

I would like to acknowledge David Rudge for providing feedback from the meeting that I could not attend, and the very careful work of John Farley and the late Gerry Geison on Pasteur's work regarding spontaneous generation and its sociopolitical context.

Epigraph: Thomas S. Hall, *Ideas of Life and Matter,* 2 vols. (Chicago: University of Chicago Press, 1969), 2:294.

1. Auguste Lutestaud, quoted in Gerald L. Geison, *The Private Science of Louis Pasteur* (Princeton, NJ: Princeton University Press, 1995), facing table of contents page.

2. Ibid., 111.

3. Summarized in John Farley, *The Spontaneous Generation Controversy from Descartes to Oparin* (Baltimore: Johns Hopkins University Press, 1974), 94–96.

4. Garland E. Allen and Jeffrey J. W. Baker, *Biology: Scientific Process and Social Issues* (New York: John Wiley and Sons, 2001), 59–61.

5. Clémence Royer, *De l'origine des Espèces par Sélection Naturelle* (Paris: Guillaumin and Masson, 1862).

6. Renan, Ernest. *Vie de Jésus* (Paris: Nelson; Calmann-Levy, 1863).

7. Félix Pouchet, *Hétérogénie; ou Traité de la Génération Spontanée* (Paris: Baillière, 1859).

8. Geison, *Private Science of Louis Pasteur,* 125–127.

9. Bruno Latour, *The Pasteurization of France* (Cambridge, MA: Harvard University Press, 1988), 8ff.

MYTH 16. THAT GREGOR MENDEL WAS A LONELY PIONEER OF GENETICS, BEING AHEAD OF HIS TIME

I am very grateful to Dick Burian, Gar Allen, and Ron Numbers for their very useful comments and suggestions on earlier versions of this chapter. I am also very grateful to Staffan Müller-Wille and Kersten Hall for letting me use their new translation of Mendel's paper.

Epigraphs: Ilona Miko, "Gregor Mendel and the Principles of Inheritance," *Nature Education* 1 (2008): 134; A. M. Winchester, *Encyclopedia Britannica Online,* s.v. "The Work of Mendel," accessed February 20, 2014, www.britannica.com/EBchecked/topic/228936/genetics/261528/The-work-of-Mendel.

1. Robert C. Olby, "Mendel No Mendelian?" *History of Science* 17 (1979): 53–72; Augustine Brannigan, "The Reification of Mendel," *Social Studies of Science* 9 (1979): 423–454.

2. For this essay I have relied on the translation of Mendel's "Versuche über Pflanzen-Hybriden" ("Experiments on Plant Hybrids") that is currently being prepared by Kersten Hall and Staffan Müller-Wille for online publication. See also Staffan Müller-Wille and Hans Jörg Rheinberger, *A Cultural History of Heredity* (Chicago: University of Chicago Press, 2012).

3. Robert C. Olby, *Origins of Mendelism,* 2nd ed. (Chicago: University of Chicago Press, 1985); Garland E. Allen, "Mendel and Modern Genetics: The Legacy for Today," *Endeavour* 27 (2003): 63–68; Sander Glibbof "The Many Sides of Gregor Mendel," in *Outsider Scientists: Routes to Innovation in Biology,* ed. Oren Harman and Michael R. Dietrich (Chicago: University of Chicago Press, 2013).

4. Olby, *Origins of Mendelism,* 100–103.

5. Mendel, "Experiments on Plant Hybrids," translation by Kersten Hall and Staffan Müller-Wille.

6. See the respective notes and textual passages in the translation of Mendel's paper by Kersten Hall and Staffan Müller-Wille.

7. Mendel, "Experiments on Plant Hybrids," translation by Kersten Hall and Staffan Müller-Wille.

8. Olby, *Origins of Mendelism,* 33 (note 4).

9. In this ratio, two classes resembled each of the parental ones (AB and ab), two classes had one character from each parental form (Ab and

aB), four classes appeared twice and had a parental and a hybrid character (ABb, aBb, AaB, Aab), and finally one class, hybrid in both characters (AaBb), appeared four times.

10. Mendel, "Experiments on Plant Hybrids," 170–171 (emphasis in the original German text, p.17, available at http://www.esp.org/foundations /genetics/classical/gm-65-f.pdf).

11. Mendel, "Experiments on Plant Hybrids," (emphasis in the original German text, p.22, available at http://www.esp.org/foundations /genetics/classical/gm-65-f.pdf).

12. Kostas Kampourakis, "Mendel and the Path to Genetics: Portraying Science as a Social Process," *Science & Education* 2 (2013): 293–324.

13. Olby, *Origins of Mendelism,* 216–220.

14. Ibid., 103–104; Brannigan, "The Reification of Mendel," 428–429.

15. Peter J. Bowler, *The Mendelian Revolution: The Emergence of Hereditarian Concepts in Modern Science and Society* (Baltimore: Johns Hopkins University Press, 1989).

16. Annie Jamieson and Gregory Radick, "Putting Mendel in His Place: How Curriculum Reform in Genetics and Counterfactual History of Science Can Work Together," in *The Philosophy of Biology: A Companion for Educators,* ed. Kostas Kampourakis (Dordrecht: Springer 2013), 577–595.

MYTH 17. THAT SOCIAL DARWINISM HAS HAD A PROFOUND INFLUENCE ON SOCIAL THOUGHT AND POLICY, ESPECIALLY IN THE UNITED STATES OF AMERICA

I thank Kostas Kampourakis and Bob Richards for their helpful suggestions.

Epigraphs: John P. Rafferty, ed., *New Thinking about Evolution* (New York: Britannica Educational Publishing, 2011), 56–57; Oswego City School District Regents Exam Prep Center (1999–2011), accessed May 9, 2015, regentsprep.org/regents/core/questions/questions.cfm?Course =ushg& TopicCode=3c.

1. Charles Darwin, *The Descent of Man and Selection in Relation to Sex,* 2 vols. (New York: D. Appleton, 1871), 1:161–162.

2. See ibid., 1:172, 2:385–386; John C. Greene, "Darwin as a Social Evolutionist," *Journal of the History of Biology* 10 (1977): 1–27; C. R. Darwin to William Graham, 3 July 1881, Darwin Correspondence Project Database (www.darwinproject.ac.uk/entry-13230); and Richard

Weikart, "A Recently Discovered Darwin Letter on Social Darwinism," *Isis* 86 (1995): 609–611.

3. Ronald L. Numbers, *Darwinism Comes to America* (Cambridge, MA: Harvard University Press, 1998), 37–38 (quoting LeConte and Powell).

4. Herbert Spencer, *Social Statics; or, The Conditions Essential to Human Happiness Specified, and the First of Them Developed* (New York: D. Appleton, 1873), 413.

5. Publisher's note in *An Autobiography,* by Herbert Spencer, 2 vols. (New York: D. Appleton, 1904), 2:113.

6. Henry Fairfield Osborn, "The Spencerian Biology," *New York Times,* April 6, 1890, 13. On Spencer's negligible influence on American religious thought, see Jon H. Roberts, *Darwinism and the Divine in America: Protestant Intellectuals and Organic Evolution, 1859–1900* (Madison: University of Wisconsin Press, 1988), 76, 274, 290.

7. Mark Francis, *Herbert Spencer and the Invention of Modern Life* (Ithaca, NY: Cornell University Press, 2007), 2.

8. William Graham Sumner, *The Forgotten Man and Other Essays,* ed. Albert Galloway Keller (New Haven, CT: Yale University Press, 1918; originally written in 1879), 225.

9. Robert C. Bannister Jr., "William Graham Sumner's Social Darwinism: A Reconsideration," *History of Political Economy* 5 (1973): 89–109, on 102. On Sumner's "un-Darwinian" approach, see Donald C. Bellomy, "'Social Darwinism' Revisited," *Perspectives in American History,* n.s., 1 (1984): 1–129; and Norman Erik Smith, "William Graham Sumner as an Anti-Social Darwinist," *Pacific Sociological Review* 22 (1979): 332–347.

10. Irvin G. Wyllie, *The Self-Made Man in America: The Myth of Rags to Riches* (New Brunswick, NJ: Rutgers University Press, 1954), 83–87.

11. Irvin G. Wyllie, "Social Darwinism and the Businessman," *Proceedings of the American Philosophical Society* 103 (1959): 629–635, on 632.

12. Charles V. Chapin, "Preventive Medicine and Natural Selection," *Journal of Social Science* 41 (1903): 54–60, at 54.

13. Ibid., 56.

14. William Jennings Bryan, *The Menace of Darwinism* (New York: Fleming H. Revell, 1921), 36–39. Bryan also quoted this passage from Darwin in his posthumously published "Last Speech," printed in *The World's Most Famous Court Trial: Tennessee Evolution Case* (Cincinnati: National Book, 1925), 335. The only earlier use of the Darwin quotation

that I have found appeared in Caleb Williams Saleeby, *Parenthood and Race Culture: An Outline of Eugenics* (London: Cassell, 1909), 171.

15. John S. Haller Jr., *Outcasts from Evolution: Scientific Attitudes of Racial Inferiority, 1859–1900* (Urbana: University of Illinois Press, 1971); Jeffrey P. Moran, "The Scopes Trial and Southern Fundamentalism in Black and White: Race, Region, and Religion," *Journal of Southern History* 70 (2004): 95–120, on 100. See also Eric D. Anderson, "Black Responses to Darwinism, 1859–1915," in *Disseminating Darwinism: The Role of Place, Race, Religion, and Gender,* ed. Ronald L. Numbers and John Stenhouse (Cambridge: Cambridge University Press, 1999), 247–266.

16. Kimberly A. Hamlin, *From Eve to Evolution: Darwin, Science, and Women's Rights in Gilded Age America* (Chicago: University of Chicago Press, 2014), 23. See also Sally Gregory Kohlstedt and Mark R. Jorgensen, "'The Irrepressible Woman Question': Women's Responses to Evolutionary Ideology," in Numbers and Stenhouse, *Disseminating Darwinism,* 266–293.

17. Mike Hawkins, *Social Darwinism in European and American Thought, 1860–1945* (Cambridge: Cambridge University Press, 1997), 97. See also Paul Crook, *Darwinism, War, and History: The Debate over the Biology of War from the "Origin of Species" to the First World War* (Cambridge: Cambridge University Press, 1994); and Edward Caudill, *Darwinian Myths: The Legends and Misuses of a Theory* (Knoxville: University of Tennessee Press, 1997), ch. 5.

18. Ronald L. Numbers, *The Creationists: From Scientific Creationism to Intelligent Design,* expanded ed. (Cambridge, MA: Harvard University Press, 2006), 55–56; David Starr Jordan, "Social Darwinism," *Public,* March 30, 1918, 400–401.

19. Bellomy, "'Social Darwinism' Revisited," 42–51, 2. See also Geoffrey M. Hodgson, "Social Darwinism in Anglophone Academic Journals: A Contribution to the History of the Term," *Journal of Historical Sociology* 17 (2004): 428–463. Hodgson (433) and others have alleged that the term "social Darwinism" first appeared in an American publication in an English translation of an article by the German zoologist Oscar Schmidt, "Science and Socialism," *Popular Science Monthly* 14 (1879): 577–591; however, this is an error.

20. D. Collin Wells, "Social Darwinism," *American Journal of Sociology* 12 (1907): 695–716, which includes Ward's commentary.

21. Lester F. Ward, "Social and Biological Struggles," *American Journal of Sociology* 13 (1907): 289–299, on 289 and 293.

22. Richard Hofstadter, *Social Darwinism in American Thought, 1860–1915* (Philadelphia: University of Pennsylvania Press, 1944), vii.

23. Ibid., 147. See also Richard Hofstadter, "William Graham Sumner, Social Darwinist," *New England Quarterly* 14 (1941): 457–477.

24. Mark A. Largent, "Social Darwinism Emerges and Is Used to Justify Imperialism, Racism, and Conservative Economic and Social Policies," in *Science and Its Times: Understanding the Social Significance of Scientific Discovery,* ed. Neil Schlager, vol. 5, *1800–1899* (Detroit: Gale Group, 2000), 134–136.

MYTH 18. THAT THE MICHELSON-MORLEY EXPERIMENT PAVED THE WAY FOR THE SPECIAL THEORY OF RELATIVITY

We would like to thank Kathy Olesko, John Heilbron, Mansoor Niaz, and the editors of this volume for helpful comments and suggestions.

Epigraph: James Richards et al., *Modern University Physics* (London: Addison Wesley, 1960), 763.

1. Gerald Holton, *Thematic Origins of Scientific Thought: Kepler to Einstein* (Cambridge, MA: Harvard University Press, 1973), 327.

2. Gerald Holton, *Thematic Origins of Scientific Thought: Kepler to Einstein,* rev. ed. (Cambridge, MA: Harvard University Press, 1988), 478.

3. See, for instance, John Stachel, *Einstein from "B" to "Z"* (Boston: Birkhäuser, 2002), 175; Arthur I. Miller, *Albert Einstein's Special Theory of Relativity: Emergence (1905) and Early Interpretation (1905–1911)* (New York: Springer, 1998), 85; and Nancy J. Nersessian, "Ad Hoc Is Not a Four-Letter Word: H. A. Lorentz and the Michelson-Morley Experiment," in *The Michelson Era in American Science,* ed. Stanley Goldberg and Roger H. Stuewer (New York: American Institute of Physics, 1988), 71–77.

4. Albert Einstein, "Über das Relativitätsprinzip und die aus Demselben Gezogenen Folgerungen," *Jahrbuch der Radioaktivität* 4 (1907): 411–462; English translation in *The Collected Papers of Albert Einstein,* vol. 2, *The Swiss Years: Writings, 1900–1909* (Princeton, NJ: Princeton University Press, 1989), 252–311.

5. See, for instance, John Norton, "Einstein's Special Theory of Relativity and the Problems in the Electrodynamics of Moving Bodies That Led Him to It," in *The Cambridge Companion to Einstein,* ed. Michel Janssen and Christoph Lehner (New York: Cambridge University Press, 2014), 72–102.

6. See Richard Staley, *Einstein's Generation: The Origins of the Relativity Revolution* (Chicago: University of Chicago Press, 2009), 11.

7. Einstein, "Über das Relativitätsprinzip," 253, 257.

8. Holton, *Thematic Origins of Scientific Thought*, rev. ed., 352.

9. Max von Laue, *Das Relativitätsprinzip* (Braunschweig: Friedrich Vieweg & Sohn, 1911), 13.

10. Albert Einstein, *Essays in Science* (New York: Philosophical Library, 1934), 49; Albert Einstein and Leopold Infeld, *The Evolution of Physics* (Cambridge: Cambridge University Press, 1938), 183–184.

11. Charles Kittel, Walter D. Knight, and Malvin A. Ruderman, *Mechanics: Berkeley Physics Course*, vol. 1, 2nd ed., revised by A. Carl Helmholz and Burton J. Moyer (New York: McGraw-Hill, 1973), 326.

12. Georg Joos with the collaboration of Ira Freeman, *Theoretical Physics*, 3rd ed. (London: Blackie and Son, 1958 [1934]), 249. See also the epigraph at the beginning of this essay.

13. See Andy Pickering, "Against Putting the Phenomena First: The Discovery of the Weak Neutral Current," *Studies in History and Philosophy of Science* 15 (1984): 85–117.

14. See Helge Kragh, "Max Planck: The Reluctant Revolutionary," *Physics World* 13 (December 2000): 31–35.

MYTH 19. THAT THE MILLIKAN OIL-DROP EXPERIMENT WAS SIMPLE AND STRAIGHTFORWARD

Epigraphs: Allvar Gullstrand, "Nobel Prize Presentation Speech," *Nobel Lectures: Physics, 1922–1941* (Amsterdam: Elsevier, 1965); *Encyclopedia Britannica Online*, s.v. "Millikan Oil-Drop Experiment," retrieved 20 March 2014, www.britannica.com/EBchecked/topic/382908/Millikan -oil-drop-experiment.

1. Joseph J. Thomson, "Cathode Rays," *Philosophical Magazine* 44 (1897): 293–316; Robert P. Crease, "Critical Point: The Most Beautiful Experiment," *Physics World* 15 (2002): 19–20.

2. Gerald Holton, "Subelectrons, Presuppositions, and the Millikan-Ehrenhaft Dispute," *Historical Studies in the Physical Sciences* 9 (1978): 161–224.

3. Robert A. Millikan, "On the Elementary Electrical Charge and the Avogadro Constant," *Physical Review* 2 (1913): 109–143.

4. Gerald Holton, "On the Hesitant Rise of Quantum Physics Research in the United States," in *The Michelson Era in American Science, 1870–1930*, ed. Stanley Goldberg and Roger H. Stuewer (New York: American Institute of Physics, 1988), 177–205.

5. Holton, "Subelectrons," 209; Gerald Holton, personal communication to the author, August 3, 2014, after having read a preliminary version of this essay (italics in the original), reproduced with permission. I owe a debt of gratitude to Holton for this contribution.

6. Ibid., 199–200, emphasis added.

7. Ibid., 184.

8. Mansoor Niaz, "The Oil Drop Experiment: A Rational Reconstruction of the Millikan-Ehrenhaft Controversy and Its Implications for Chemistry Textbooks," *Journal of Research in Science Teaching* 37 (2000): 480–508; María A. Rodríguez and Mansoor Niaz, "The Oil Drop Experiment: An Illustration of Scientific Research Methodology and Its Implications for Physics Textbooks," *Instructional Science* 32 (2004): 357–386.

9. Holton, "Subelectrons"; Mansoor Niaz, "An Appraisal of the Controversial Nature of the Oil Drop Experiment: Is Closure Possible?" *British Journal for the Philosophy of Science* 56 (2005): 681–702. This article reviews various interpretations of the oil-drop experiment and attempts closure.

10. Rodríguez and Niaz, "The Oil Drop Experiment," 375–377; Stephen Klassen, "Identifying and Addressing Student Difficulties with the Millikan Oil Drop Experiment," *Science & Education* 18 (2009): 593–607.

MYTH 20. THAT NEO-DARWINISM DEFINES EVOLUTION AS RANDOM MUTATION PLUS NATURAL SELECTION

My thanks to Dick Burian and the editors for helping me improve this chapter.

Epigraph: Christoph Schönborn, "Finding Design in Nature," *New York Times,* July 7, 2005, A27.

1. Julian Huxley, *Evolution: The Modern Synthesis* (London: Allen and Unwin, 1942).

2. A Scientific Dissent from Darwinism, Discovery Institute, accessed April 13, 2014, www.dissentfromdarwin.org.

3. Pius XII, *Humani Generis,* 1950, at http://w2.vatican.va/content/pius-xii/en/encyclicals/documents/hf_p-xii_enc_12081950_humani-generis.html; John Paul II, "Message to the Pontifical Academy of Sciences: On Evolution," 22 October 1996, www.ewtn.com/library/PAPALDOC/JP961022.HTM.

4. G. M. Auletta, M. Leclerc, and R. Martinez, eds., *Biological Evolution: Facts and Theories; A Critical Appraisal 150 Years after "The*

Origin of Species" (Rome: Gregorian and Biblical Press, 2011); Ronald L. Numbers, *The Creationists: From Scientific Creationism to Intelligent Design,* expanded ed. (Cambridge, MA: Harvard University Press, 2006), 395–396.

5. August Weismann, *The Germ-Plasm: A Theory of Heredity* (New York: Charles Scribner's Sons, 1893 [1892]), xiii.

6. Wilhelm Johannsen, *Elemente der exakten Erblichkeitslehre* (Jena: Gustav Fischer, 1909).

7. Talk about factors of evolution and contestation about which is the "creative" factor go back to Herbert Spencer, *The Factors of Organic Evolution* (New York: D. Appleton, 1887). Factors are sometimes also referred to as agents or forces of evolution.

8. Vernon Kellogg, *Darwinism To-Day: A Discussion of Present-Day Scientific Criticism of the Darwinian Selection Theories* (New York: Holt, 1907).

9. William Provine, *The Origins of Theoretical Population Genetics* (Chicago: University of Chicago Press, 1971).

10. The theorem is the Hardy-Weinberg equilibrium formula, named after the mathematicians who simultaneously but independently derived it. See G. Hardy, "Mendelian Proportions in a Mixed Population," *Science* 28 (1908): 49–50. The reconciliation of Mendelism and Darwinism was first set out in Ronald A. Fisher, "The Correlation of Relations on the Supposition of Mendelian Inheritance," *Transactions of the Royal Society of Edinburgh* 52 (1918): 399–433.

11. Huxley, *Evolution,* 28; Theodosius Dobzhansky, *Genetics and the Evolutionary Process* (New York: Columbia University Press, 1970), 430–431; Ernst Mayr, introduction to *The Evolutionary Synthesis,* by Ernst Mayr and William Provine, eds. (Cambridge, MA: Harvard University Press, 1980), 18, 22.

12. On genetic drift, see Roberta Millstein, Robert Skipper, and Michael Dietrich, "(Mis)interpreting Mathematical Models: Drift as a Physical Process," *Philosophy and Theory in Biology* 1 (2009): e002.

13. Theodosius Dobzhansky, remark at the University of Chicago Darwin Centennial Celebration, November 1959, Panel 2, in Sol Tax and Charles Callendar, eds., *Evolution after Darwin,* 3 vols. (Chicago: University of Chicago Press, 1960), 3:115.

14. Sean B. Carroll, *Endless Forms Most Beautiful: The New Science of Evo Devo and the Making of the Animal Kingdom* (New York: W. W. Norton, 2005); Scott Gilbert and David Epel, *Ecological Developmental Biology* (Sunderland, MA: Sinauer, 2009).

15. Eva Jablonka and Marian Lamb, *Evolution in Four Dimensions: Genetic, Epigenetic, Behavioral, and Symbolic Variation in the History of Life* (Cambridge, MA: MIT Press, 2005).

16. David Depew, "Conceptual Change and the Rhetoric of Evolutionary Theory: 'Force Talk' as a Case Study and Challenge for Science Pedagogy," in *The Philosophy of Biology: A Companion for Educators,* ed. Kostas Kampourakis (Dordrecht: Springer, 2013), 121–144.

17. Jonathan Wells, "Evolution and Intelligent Design," Discovery Institute, June 1, 1997, www.discovery.org/a/77.

18. Stephen Jay Gould, *Wonderful Life: The Burgess Shale and the Nature of History* (New York: W. W. Norton, 1989), stresses accident in the larger history of life on earth. Simon Conway Morris, *The Crucible of Creation: The Burgess Shale and the Rise of Animals* (Oxford: Oxford University Press, 1998); and Simon Conway Morris, *Life's Solution: Inevitable Humans in a Lonely Universe* (Cambridge: Cambridge University Press, 2003), stress cumulative adaptedness.

MYTH 21. THAT MELANISM IN PEPPERED MOTHS IS NOT A GENUINE EXAMPLE OF EVOLUTION BY NATURAL SELECTION

The author thanks Bruce Grant for looking over an earlier version of this chapter, David Depew for his additional suggestions, and especially Kostas Kampourakis and Ron Numbers for their advice.

Epigraph: George B. Johnson and Jonathan B. Losos, *The Living World,* 6th ed. (New York: McGraw-Hill, 2010), 303.

1. John B. S. Haldane, "A Mathematical Theory of Natural and Artificial Selection," *Transactions of the Cambridge Philosophical Society* 23 (1924): 19–41.

2. Edmund B. Ford, "Genetic Research in the Lepidoptera," *Annals of Eugenics* 10 (1940): 227–252.

3. Henry B. D. Kettlewell, "Selection Experiments on Industrial Melanism in the Lepidoptera," *Heredity* 9 (1955): 323–342; Henry B. D. Kettlewell, "Further Selection Experiments on Industrial Melanism in the Lepidoptera," *Heredity* 10 (1956): 287–301.

4. David W. Rudge and Janice M. Fulford, "The Role of Visual Imagery in Textbook Portrayals of Industrial Melanism," in *Science and Culture: Promise, Challenge and Demand; Book of Proceedings for the Eleventh International History, Philosophy and Science Teaching (IHPST) and Sixth Greek History, Philosophy and Science Teaching Joint Conference,* ed. Fanny Seroglou, Vassilis Koulountzos, and Anastasios Siatras (Thessaloniki, Greece: Aristotle University, 2011), 630–637.

5. Bruce S. Grant, "Fine Tuning the Peppered Moth Paradigm," *Evolution* 53 (1999): 980–984; Lawrence M. Cook, "The Rise and Fall of the Carbonaria Form of the Peppered Moth," *Quarterly Review of Biology* 78 (2003): 1–19; Bruce S. Grant et al., "Geographic and Temporal Variation in the Incidence of Melanism in Peppered Moth Populations in America and Britain," *Journal of Heredity* 89 (1998): 465–471; Lawrence M. Cook et al., "Selective Bird Predation on the Peppered Moth: The Last Experiment of Michael Majerus," *Biological Letters* 8 (2012): 609–612; Michael E. N. Majerus, *Melanism: Evolution in Action* (Oxford: Oxford University Press, 1998).

6. Jonathan Wells, *Icons of Evolution: Science or Myth? Why Much of What We Teach about Evolution Is Wrong* (Washington, DC: Regnery Publishing, 2002), x. For a critical appraisal, see David W. Rudge, "Cryptic Designs on the Peppered Moth," *International Journal of Tropical Biology and Conservation* 50 (2002): 1–7.

7. Judith Hooper, *Of Moths and Men: An Evolutionary Tale: Intrigue, Tragedy and the Peppered Moth* (London: Fourth Estate, 2002). For a critical appraisal, see David W. Rudge, "Did Kettlewell Commit Fraud? Re-examining the Evidence," *Public Understanding of Science* 14 (2005): 249–268.

8. Rudge and Fulford, "Role of Visual Imagery," 632.

9. David W. Rudge, "H. B. D. Kettlewell's Research, 1937–1953: The Influence of E. B. Ford, E. A. Cockayne and P. M. Sheppard," *History and Philosophy of the Life Sciences* 28 (2006): 359–388.

10. David W. Rudge, "Does Being Wrong Make Kettlewell Wrong for Science Teaching?" *Journal of Biology Education* 35 (2000): 5–11.

MYTH 22. THAT LINUS PAULING'S DISCOVERY OF THE MOLECULAR BASIS OF SICKLE-CELL ANEMIA REVOLUTIONIZED MEDICAL PRACTICE

I thank Kostas Kampourakis, Ron Numbers, and Peter Ramberg for their helpful comments.

Epigraphs: Alan N. Schechter and Griffin P. Rodgers, "Sickle Cell Anemia: Basic Research Reaches the Clinic," *New England Journal of Medicine* 20 (1995): 1372–1374, on 1372; Teresa Audesirk, Gerald Audesirk, and Bruce E. Byers, *Biology: Life on Earth*, 6th ed. (San Francisco: Benjamin Cummings, 2011), 227–228.

1. Dorothy Nelkin and Suzan Lindee, *The DNA Mystique: The Gene as a Cultural Icon* (Ann Arbor: University of Michigan Press, 2004);

Sheldon Krimsky and Jeremy Gruber, eds., *Genetic Explanations: Sense and Nonsense* (Cambridge, MA: Harvard University Press, 2013).

2. Thomas Hager, *Force of Nature: The Life of Linus Pauling* (New York: Simon and Schuster, 1995); Linus Pauling et al., "Sickle Cell Anemia, a Molecular Disease," *Science* 110 (1949): 543–548; Bruno J. Strasser, "Sickle Cell Anemia, a Molecular Disease," *Science* 286 (1999): 1488–1490; Bruno J. Strasser, "Linus Pauling's 'Molecular Diseases': Between History and Memory," *American Journal of Medical Genetics* 115 (2002): 83–93; Teresa Audesirk, Gerald Audesirk, and Bruce E. Byers, *Biology: Life on Earth* (San Francisco: Benjamin Cummings, 2011), 227.

3. Roland Barthes, *Mythologies* (Paris: Editions du Seuil, 1957); Bruno J. Strasser, "Who Cares about the Double Helix?" *Nature* 422 (2003): 803–804; Pnina G. Abir-Am and Clark A. Elliott, eds., *Commemorative Practices in Science: Historical Perspectives on the Politics of Collective Memory* (Chicago: University of Chicago Press, 2000).

4. Schechter and Rodgers, "Sickle Cell Anemia," 1372; I. M. Klotz, D. N. Haney, and L. C. King, "Rational Approaches to Chemotherapy: Antisickling Agents," *Science* 213 (1981): 724–731.

5. Soraya de Chadarevian and Harmke Kamminga, eds., *Molecularizing Biology and Medicine: New Practices and Alliances, 1910s–1970s* (Amsterdam: Harwood Academic Publishers, 1998); Pauling, "Sickle Cell Anemia," 547.

6. John Parascandola, "The Theoretical Basis of Paul Ehrlich's Chemotherapy," *Journal of the History of Medicine and Allied Sciences* 36 (1981): 19–43; Robert Bud, *Penicillin: Triumph and Tragedy* (Oxford: Oxford University Press, 2007); Jack E. Lesch, *The First Miracle Drugs: How the Sulfa Drugs Transformed Medicine* (New York: Oxford University Press, 2007); Jordan Goodman and Vivien Walsh, *The Story of Taxol: Nature and Politics in the Pursuit of an Anti-Cancer Drug* (Cambridge: Cambridge University Press, 2001).

7. Strasser, "Linus Pauling."

8. Lily E. Kay, *The Molecular Vision of Life: Caltech, the Rockefeller Foundation, and the Rise of the New Biology* (New York: Oxford University Press, 1993), ch. 6; Suzan M. Lindee, *Moments of Truth in Genetic Medicine* (Baltimore: Johns Hopkins University Press, 2005); Nathaniel C. Comfort, *The Science of Human Perfection: How Genes Became the Heart of American Medicine* (New Haven, CT: Yale University Press, 2012).

9. Schechter and Rodgers, "Sickle Cell Anemia"; Keith Wailoo and Stephen G. Pemberton, *The Troubled Dream of Genetic Medicine: Ethnicity and Innovation in Tay-Sachs, Cystic Fibrosis, and Sickle Cell Dis-*

ease (Baltimore: Johns Hopkins University Press, 2006); Bruno J. Strasser, "Response," *Science* 287 (2000): 593; Valentine Brousse, Julie Makani, and David C. Rees, "Management of Sickle Cell Disease in the Community," *British Medical Journal* 348 (2014): 1765–1789.

10. Simon D. Feldman and Alfred I. Tauber, "Sickle Cell Anemia: Re-examining the First 'Molecular Disease,' " *Bulletin of the History of Medicine* 71 (1997): 623–650.

11. Linus Pauling to J. B. S. Haldane, July 18, 1955, Eva Helen and Linus Pauling Papers, Oregon State University; Linus Pauling, "The Hemoglobin Molecule in Health and Disease," *Proceedings of the American Philosophical Society,* 96 (1952): 556–565, on 564; Linus Pauling, typescript of "Abnormal Hemoglobin Molecules in Relation to Disease," 1956, Pauling Papers, 22.

12. Kay, *Molecular Vision of Life;* Jean-Paul Gaudillière, *Inventer la Biomédecine: La France, l'Amérique et la Production des Savoirs du Vivant, 1945–1965* (Paris: La Découverte, 2002); Bruno J. Strasser, "Institutionalizing Molecular Biology in Post-War Europe: A Comparative Study," *Studies in the History and Philosophy of Biological and Biomedical Sciences* 33C (2002): 533–564.

13. Vivian Quirke and Jean-Paul Gaudillière, "The Era of Biomedicine: Science, Medicine, and Public Health in Britain and France after the Second World War," *Medical History* 52 (2008): 441–452; Bruno J. Strasser, "Magic Bullets and Wonder Pills: Making Drugs and Diseases in the Twentieth Century," *Historical Studies in the Natural Sciences* 38 (2008): 303–312; Bruno J. Strasser, *Biomedicine: Meanings, Assumptions, and Possible Futures* (Berne: CSST, 2014).

14. Evelyn F. Keller, *The Mirage of a Space between Nature and Nurture* (Durham, NC: Duke University Press, 2010).

MYTH 23. THAT THE SOVIET LAUNCH OF *SPUTNIK* CAUSED THE REVAMPING OF AMERICAN SCIENCE EDUCATION

Epigraphs: Larry Abramson, "*Sputnik* Left Legacy for U.S. Science Education," September 30, 2007, www.npr.org/templates/story/story .php?storyId=14829195; Barack H. Obama, State of the Union Address, January 25, 2011, www.whitehouse.gov/the-press-office/2011/01/25 /remarks-president-state-union-address.

1. Paul Dickson, *Sputnik: The Shock of the Century* (New York: Walker Publishing, 2001). The Eisenhower quotation appears in Robert A. Divine, *The Sputnik Challenge: Eisenhower's Response to the Soviet Satellite* (New York: Oxford University Press, 1993), 15.

2. For details on the PSSC curriculum project, see John L. Rudolph, *Scientists in the Classroom: The Cold War Reconstruction of American Science Education* (New York: Palgrave Macmillan, 2002), chs. 4 and 5. The Zacharias quotation appears in an interview with Zacharias in the PSSC Oral History Collection, Institute Archives and Special Collections, Massachusetts Institute of Technology, Cambridge, Massachusetts.

3. AIBS Executive Committee meeting minutes, January 4, 1952, Papers of the American Institute of Biological Sciences, AIBS, Washington, DC, box 2; NRC, Conference on Biological Education meeting minutes, March 10, 1953, National Academy of Sciences, Committee on Educational Policies [NAS/CEP], National Academy of Sciences Archives, Washington, DC. That summer workshop resulted in Chester A. Lawson and Richard E. Paulson, *Laboratory and Field Studies in Biology: A Sourcebook for Secondary Schools* (New York: Holt, Rinehart and Winston, 1958). For an overview of the origin of BSCS, see Rudolph, *Scientists in the Classroom*, ch. 6.

4. J. Merton England, *A Patron for Pure Science: The National Science Foundation's Formative Years, 1945–57* (Washington, DC: National Science Foundation, 1982); Daniel Lee Kleinman, *Politics on the Endless Frontier* (Durham, NC: Duke University Press, 1995). On the origin of the chemistry project, see Paul Westmeyer, "The Chemical Bond Approach to Introductory Chemistry," *School Science and Mathematics* 61 (1961): 317–322.

5. Diane Ravitch, *The Troubled Crusade: American Education, 1945–1980* (New York: Basic Books, 1983), 26–41; Carl F. Kaestle, "Federal Aid to Education since World War II: Purposes and Politics," in *The Future of the Federal Role in Education,* ed. Jack Jennings (Washington, DC: Center on Education and Policy, 2001), 13–36.

6. Hillier Krieghbaum and Hugh Rawson, *An Investment in Knowledge: The First Dozen Years of the National Science Foundation's Summer Institutes Programs to Improve Secondary School Science and Mathematics Teaching, 1954–1965* (New York: New York University Press, 1969), 65–82.

7. Nicholas DeWitt, *Soviet Professional Manpower: Its Education, Training, and Supply* (Washington, DC: National Science Foundation, 1955).

8. The Thomas quotation appears in United States House of Representatives, *Hearings before the Subcommittee on Independent Offices,* January 30, 1956 (Washington, DC: U.S. Government Printing Office, 1956), 528, 522. Detonation of the deliverable device occurred in No-

vember 1955. The Soviet Union's first successful explosion of a stationary
hydrogen bomb had occurred in August 1953; see Walter A. McDou-
gall, *The Heavens and the Earth: A Political History of the Space Age*
(New York: Basic Books, 1985), 55. On the National Defense Education Act, see Wayne J. Urban, *More Than Science and Sputnik: The
National Defense Education Act of 1958* (Tuscaloosa: University of
Alabama Press, 2010).

9. Among the main critiques of public education at this time were
Bernard Iddings Bell, *Crisis in Education: A Challenge to American Com-
placency* (New York: Whittlesey House, 1949); Albert Lynd, *Quackery in
the Public Schools* (Boston: Little, Brown, 1953); Paul Woodring, *Let's
Talk Sense about Our Schools* (New York: McGraw-Hill, 1953); and
Arthur E. Bestor, *Educational Wastelands: The Retreat from Learning in
Our Public Schools* (Urbana: University of Illinois Press, 1953). The
crisis atmosphere surrounding American education is detailed further in
Rudolph, *Scientists in the Classroom*, ch. 1; and Ravitch, *Troubled Crusade*, ch. 1.

MYTH 24. THAT RELIGION HAS TYPICALLY IMPEDED THE PROGRESS OF SCIENCE

Epigraphs: Jerry A. Coyne, "Science and Religion Aren't Friends," *USA
Today,* October 11, 2010, http://usatoday30.usatoday.com/news/opinion
/forum/2010-10-11-column11_ST_N.htm.
Sam Harris, "Science Must Destroy Religion," *Huffington Post,* January
2, 2006, reprinted in *What Is Your Dangerous Idea?,* ed. John Brockman
(New York: Harper, 2007), 148–151.

1. For debunking of many of these myths, see Ronald L. Numbers,
ed., *Galileo Goes to Jail and Other Myths about Science and Religion*
(Cambridge, MA: Harvard University Press, 2009). See also Peter Har-
rison, ed., *The Cambridge Companion to Science and Religion,* chs. 1–5;
Jon H. Roberts, "Science and Religion," in *Wrestling with Nature: From
Omens to Science,* ed. Peter Harrison, Ronald L. Numbers, and Mi-
chael H. Shank (Chicago: University of Chicago Press, 2011), 253–279.
More generally, see John Brooke, *Science and Religion: Some Historical
Perspectives* (Cambridge: Cambridge University Press, 1991).

2. John Heilbron, *The Sun in the Church: Cathedrals as Solar Obse-
vatories* (Cambridge, MA: Harvard University Press, 1999), 3.

3. David Livingstone, *Darwin's Forgotten Defenders: The Encounter
between Evangelical Theology and Evolutionary Thought* (Vancouver:
Regent College Publishing, 1984); Jon H. Roberts, *Darwinism and the*

Divine in America, 2nd ed. (Notre Dame, IN: University of Notre Dame Press, 2001).

4. Ronald L. Numbers, *The Creationists: From Scientific Creationism to Intelligent Design,* expanded ed. (Cambridge, MA: Harvard University Press, 2006).

5. Michael H Shank, "That the Medieval Christian Church Suppressed the Growth of Science," in Numbers, *Galileo Goes to Jail,* 19–27, on 21–22.

6. See, for example, Brooke, *Science and Religion;* David C. Lindberg and Ronald L. Numbers, eds., *When Science and Christianity Meet* (Chicago: University of Chicago Press, 2008); Gary Ferngren, ed., *Science and Religion: An Historical Introduction* (Baltimore: Johns Hopkins University Press, 2002); Peter Harrison, *The Fall of Man and the Foundations of Science* (Cambridge: Cambridge University Press, 2007); Peter Harrison, "Laws of Nature in Seventeenth-Century England," in *The Divine Order, the Human Order, and the Order of Nature,* ed. Eric Watkins (New York: Oxford University Press, 2013), 127–148; and Stephen Gaukroger, *The Emergence of a Scientific Culture* (Oxford: Oxford University Press, 2006).

7. Peter Harrison, *The Territories of Science and Religion* (Chicago: University of Chicago Press, 2015).

8. Cotton Mather, *American Tears upon the Ruines of the Greek Churches* (Boston, 1701), 42–43.

9. John Milton, *Areopagitica* (Indianapolis: Liberty Fund, 1999; first published in 1644), 31f.

10. Voltaire, "Newton and Descartes," *Philosophical Dictionary,* 2nd ed., 6 vols. (London: John and Henry Hunt, 1824), 5:113.

11. Jean Le Rond d'Alembert, *Preliminary Discourse to the Encyclopedia of Diderot,* trans. Richard N. Schwab and Walter E. Rex (Chicago: University of Chicago Press, 1995), 74.

12. Nicolas de Condorcet, *Sketch for a Historical Picture of the Progress of the Human Mind,* trans. June Barraclough (New York: Noonday, 1955), 72.

13. For references to Comte and Buckle, see Friedrich Albert Lange, *History of Materialism and Critique of Its Present Significance,* 2nd ed., trans. Ernest Chester Thomas, 3 vols. (Boston: James Osgood, 1877), 1:4. See also William Whewell, *History of the Inductive Sciences from the Earliest to the Present Time,* 2 vols. (New York: D. Appleton, 1858), 1:255.

14. John William Draper, *History of the Conflict between Science and Religion* (New York: D. Appleton, 1874), 52, 157–159, 160–161, 168–

169; Andrew Dickson White, A *History of the Warfare of Science with Theology in Christendom,* 2 vols. (New York: D. Appleton, 1896), 1:71–74, 108, 118; 2:49–55, 55–63. For a debunking of many of these myths, see Numbers, *Galileo Goes to Jail.*

15. Pippa Norris and Ronald Inglehart, *Sacred and Secular: Religion and Politics Worldwide* (Cambridge: Cambridge University Press, 2004), 68.

MYTH 25. THAT SCIENCE HAS BEEN LARGELY A SOLITARY ENTERPRISE

I wish to express my gratitude to the participants at the conference on Newton's Apple and Other Historical Myths about Science for lively discussions and to the editors of this volume for their perceptive comments on this essay.

Epigraphs: Edmund Turnor, *Collections for the History of the Town and Soke of Grantham: Containing Authentic Memoirs of Sir Isaac Newton* (London: William Miller, 1806), 173n2; Decca Aitkenhead, "Peter Higgs: I Wouldn't Be Productive Enough for Today's Academic System," *Guardian,* December 6, 2013, www.theguardian.com/science/2013/dec/06/peter-higgs-boson-academic-system.

1. Joseph Warton and Others, eds., *The Works of Alexander Pope,* 9 vols. (London: Richard Priestley, 1822), 2:379.

2. Frank E. Manuel, *A Portrait of Isaac Newton* (Cambridge, MA: Harvard University Press, 1968); Robert Palter, ed. *The Annus Mirabilis of Isaac Newton* (Cambridge, MA: MIT Press, 1971); Richard S. Westfall, *Never at Rest: A Biography of Sir Isaac Newton* (Cambridge: Cambridge University Press, 1980). For further elaboration, see Myth 6 in this volume.

3. William H. McNeill, "Mythistory, or Truth, Myth, History, and Historians," *American Historical Review* 91 (1986): 1–10.

4. David Park, *Introduction to Quantum Theory* (New York: McGraw Hill, 1964), 2–4; Theresa Levitt, *A Short Bright Flash: Augustin Fresnel and the Birth of the Modern Lighthouse* (New York: W. W. Norton, 2013).

5. Charlotte Abney Salomon, "Finding Yttrium: Johan Gadolin and the Development of a 'Discovery,'" presented at a conference on Scientific Revolutions, Rutgers University, February 28, 2014. A similar story could be told about the textbook treatment of Gregor Mendel's work; for details, see Kostas Kampourakis, "Mendel and the Path to Genetics:

Portraying Science as a Social Process," *Science & Education* 22 (2013): 293–324.

6. Ola Halldén, "Conceptual Change and the Learning of History," *International Journal of Education Research* 27 (1997): 201–210; Herbert Butterfield, *The Origins of Modern Science, 1300–1800* (London: G. Bell, 1949); Charles Coulston Gillispie, *The Edge of Objectivity: An Essay in the History of Scientific Ideas* (Princeton, NJ: Princeton University Press, 1960).

7. Thomas S. Kuhn, *The Structure of Scientific Revolutions* (Chicago: University of Chicago Press, 1962); Ludwik Fleck, *Genesis and Development of a Scientific Fact*, ed. Thaddeus J. Trenn and Robert K. Merton, trans. Frederick Bradley (Chicago: University of Chicago Press, 1979); Steven Shapin and Simon Schaffer, *Leviathan and the Air Pump: Hobbes, Boyle, and the Experimental Life* (Princeton, NJ: Princeton University Press, 1985); Bruno Latour, *Laboratory Life: The Construction of Scientific Facts* (Princeton, NJ: Princeton University Press, 1986); Jan Golinski, *Making Natural Knowledge: Constructivism and the History of Science* (Cambridge: Cambridge University Press, 1998).

8. Steven Shapin, "The Mind Is Its Own Place: Science and Solitude in Seventeenth-Century England," *Science in Context* 4 (1991): 191–218, on 211.

9. Patricia Fara, "Isaac Newton Lived Here: Sites of Memory and Scientific Heritage," *British Journal for the History of Science* 33 (2000): 407–426; Patricia Fara, *Newton: The Meaning of Genius* (London: Macmillan, 2002); Simon Schaffer, "Newton on the Beach: The Information Order of the *Principia Mathematica*," *History of Science* 47 (2009): 243–276; Jim Endersby, "Editor's Introduction," in Charles Darwin, *On the Origin of Species by Means of Natural Selection; or, The Preservation of Favoured Races in the Struggle for Life*, ed. Jim Endersby (Cambridge: Cambridge University Press, 2009), xi–lxv.

10. David DeVorkin, *Henry Norris Russell: Dean of American Astronomers* (Princeton, NJ: Princeton University Press, 2000), 216.

11. Naomi Oreskes, "Objectivity or Heroism: On the Invisibility of Women in Science," *Osiris* 11 (1996): 87–113; Steven Shapin, "The Invisible Technician," *American Scientist* 17 (1989): 554–563; Margaret W. Rossiter, *Women Scientists in America*, 3 vols. (Baltimore: Johns Hopkins University Press, 1982–2012).

12. See Intergovernmental Panel on Climate Change, www.ipcc.ch.

MYTH 26. THAT THE SCIENTIFIC METHOD ACCURATELY REFLECTS WHAT SCIENTISTS ACTUALLY DO

Epigraph: Wikipedia, s.v. "Scientific Method," accessed November 29, 2014, http://en.wikipedia.org/wiki/Scientific_method.

1. See, for instance, Henry H. Bauer, *Scientific Literacy and the Myth of the Scientific Method* (Urbana-Champaign: University of Illinois Press, 1994).

2. "How Science Works: The Flowchart," Understanding Science: How Science *Really* Works, undsci.berkeley.edu/article/scienceflowchart. This website was created through a partnership between the University of California Museum of Paleontology and the University of California at Berkeley and funded by the National Science Foundation.

3. John A. Schuster and Richard R. Yeo, eds., *The Politics and Rhetoric of Scientific Method: Historical Studies* (Dordrecht: D. Reidel, 1986).

4. For the definitive account of "scientific method" in education, see John L. Rudolph, "Epistemology for the Masses: The Origins of the 'Scientific Method' in Education," *History of Education Quarterly* 45 (2005): 341–376.

5. Raymond Williams, *Keywords: A Vocabulary of Culture and Society* (New York: Oxford University Press, 1976), 17.

6. Stanley Jevons, *Principles of Science* (London: Macmillan, 1874), vii; Stuart Rice, introduction to *Methods in Social Science*, ed. Stuart Rice (Chicago: University of Chicago Press, 1931), 5.

7. G. Nigel Gilbert and Michael Mulkay, "Warranting Scientific Belief," *Social Studies of Science* 12 (1982): 383–408.

8. For a general methodological overview, see Laurens Laudan, "Theories of Scientific Method from Plato to Mach," *History of Science* 7 (1968): 1–63; and Barry Gower, *Scientific Method* (London: Routledge, 1997). On the Scientific Revolution, see Steven Shapin, *The Scientific Revolution* (Chicago: University of Chicago Press, 1996).

9. The three databases used here include the catalog of the Library of Congress (http://catalog.loc.gov), the *American Periodicals Series Online* (http://www.proquest.com/products_pq/descriptions/aps.shtml), and the catalog of the contents of the *New York Times* offered by *Proquest Historical Newspapers* (http://www.proquest.com/products_pq/descriptions/pq-hist-news.shtml). I performed all these searches between May and June 2004. The first was on "scientific method" in book titles. The second two were on "scientific method" in the full text of articles. In all cases, I divided by the total number of items about science to gain a better image

of the relative prominence of the scientific method as an aspect of science talk.

10. These conclusions are drawn from the same set of databases, in addition to other important catalogs, including 19th Century Masterfile (www.paratext.com/19th-century-masterfile); Making of America (www .hti.umich.edu/m/moagrp and http://cdl.library.cornell.edu/moa); and *Readers' Guide* (www.hwwilson.com/databases/Readersg.htm), using the same kinds of searches mentioned in the previous note. See also Williams, *Keywords*.

11. See, for instance, Daniel Patrick Thurs, *Science Talk* (New Brunswick, NJ: Rutgers University Press, 2007); and Peter Harrison, Ronald L. Numbers, and Michael H. Shank, eds., *Wrestling with Nature: From Omens to Science* (Chicago: University of Chicago Press, 2011).

12. See Thomas F. Gieryn, *Cultural Boundaries of Science: Credibility on the Line* (Chicago: University of Chicago Press, 1999).

13. "The Future of Human Character," *Ladies' Repository,* January 1868, 43.

14. Quoted in Daniel J. Kevles, *The Physicists: The History of a Scientific Community in Modern America* (New York: Alfred A. Knopf, 1978), 98.

15. W. C. Croxton, *Science in the Elementary School* (New York: McGraw-Hill, 1937), 337.

16. Robert A. Millikan, "The Diffusion of Science: The Natural Sciences," *Scientific Monthly* 35 (1932): 205.

17. See Rudolph, "Epistemology for the Masses."

18. John Dewey, quoted in Louise Nichols, "The High School Student and Scientific Method," *Journal of Educational Psychology* 20 (March 1929): 196; Nelson B. Henry, ed., *46th Yearbook of the National Society for the Study of Education* (Chicago: University of Chicago Press, 1947), 62.

19. John B. Watson, "What Is Behaviorism?" *Harper's Monthly* 152 (1926): 724; Dorothy Ross, *The Origins of American Social Science* (Cambridge: Cambridge University Press, 1991), 401–402.

20. Michael Schudson, *Discovering the News: A Social History of American Newspapers* (New York: Basic Books, 1978), 7–8; George Gallup, "A Scientific Method for Determining Reader-Interest," *Journalism Quarterly* 7 (1930): 1–13.

21. David Hollinger, "Justification by Verification: The Scientific Challenge to the Moral Authority of Christianity in Modern America," in *Religion and Twentieth-Century American Intellectual Life,* ed. Michael H. Lacey (Cambridge: Cambridge University Press, 1989), 116–135.

22. Paul Feyerabend, *Against Method* (London: Humanties Press, 1975). See also Paul Feyerabend, *Killing Time* (Chicago: University of Chicago Press, 1995).

23. Helen P. Libel, "History and the Limitations of Scientific Method," *University of Toronto Quarterly* 34 (October 1964): 15–16; Walter A. Thurber and Alfred T. Collette, *Teaching Science in Today's Secondary Schools*, 2d ed. (Boston: Allyn and Bacon, 1964), 7. In general, see John L. Rudolph, *Scientists in the Classroom: The Cold War Reconstruction of American Science Education* (New York: Palgrave Macmillan, 2002).

MYTH 27. THAT A CLEAR LINE OF DEMARCATION HAS SEPARATED SCIENCE FROM PSEUDOSCIENCE

The author would like to thank Kostas Kampourakis, Erika Milam, Ron Numbers, and Michael Ruse for helpful comments on an earlier version of this essay.

Epigraph: Paul G. Hewitt, *Conceptual Physics: The High School Physics Program* (Needham, MA: Prentice Hall, 2002), 4.

1. Chris Mooney and Sheril Kirshenbaum, *Unscientific America: How Scientific Illiteracy Threatens Our Future* (New York: Basic Books, 2009).

2. "The Sacred Disease," in *The Medical Works of Hippocrates*, ed. and trans. John Chadwick and W. N. Mann (Oxford: Blackwell Scientific Publications, 1950), 179–193, on 179. We no longer believe the author's wind-and-brain etiology but do endorse his critique of theological causation.

3. Thomas Nickles, "The Problem of Demarcation: History and Future," in *Philosophy of Pseudoscience: Reconsidering the Demarcation Problem*, ed. Massimo Pigliucci and Maarten Boudry (Chicago: University of Chicago Press, 2013): 101–120.

4. For a discussion of further conceptual difficulties of the problem, see Martin Mahner, "Science and Pseudoscience: How to Demarcate after the (Alleged) Demise of the Demarcation Problem," in Pigliucci and Boudry, *Philosophy of Pseudoscience*, 29–43, on 31–33.

5. A good survey of many fringe topics is Daniel Patrick Thurs and Ronald L. Numbers, "Science, Pseudo-Science, and Science Falsely So-Called," in *Wrestling with Nature: From Omens to Science*, ed. Peter Harrison, Ronald L. Numbers, and Michael H. Shank (Chicago: University of Chicago Press, 2011), 281–305.

6. Karl Popper, "Science: Conjectures and Refutations," in Popper, *Conjectures and Refutations: The Growth of Scientific Knowledge*

(New York: Routledge, 2002 [1963]), 43–78, on 44. Emphasis in original.

7. Ibid., 47–48. Emphasis in original.

8. It was originally published as Karl Popper, "Philosophy of Science: A Personal Report," *British Philosophy in Mid-Century*, ed. C. A. Mace (London: George Allen and Unwin, 1957), 155–189.

9. Larry Laudan, "The Demise of the Demarcation Problem," in *But Is It Science?: The Philosophical Question in the Creation/Evolution Controversy*, ed. Michael Ruse, updated edition (Amherst, NY: Prometheus Books, 1988), 337–350, on 346.

10. Massimo Pigliucci, "The Demarcation Problem: A (Belated) Response to Laudan," in Pigliucci and Boudry, *Philosophy of Pseudoscience*, 9–28.

11. Judge William R. Overton, "United States District Court Opinion: *McLean v. Arkansas Board of Education*," in Ruse, *But Is It Science?*, 307–331, on 318. For the relevant section of the Ruse testimony, see "Witness Testimony Sheet: *McLean v. Arkansas Board of Education*," in ibid., 287–306, on 300–304. See also Edward J. Larson and Ronald L. Numbers, "Creation, Evolution, and the Boundaries of Science: The Debate in the United States," *Almagest: International Journal for the History of Scientific Ideas* 3 (May 2012): 4–24.

12. Judge John E. Jones II in *Tammy Kitzmiller, et al. v. Dover Area School District, et al.*, 400 F. Supp. 2d 707 (M.D. Pa. 2005).

13. Michael D. Gordin, *The Pseudoscience Wars: Immanuel Velikovsky and the Birth of the Modern Fringe* (Chicago: University of Chicago Press, 2012), 202.

14. Naomi Oreskes and Erik M. Conway, *Merchants of Doubt: How a Handful of Scientists Obscured the Truth on Issues from Tobacco Smoke to Global Warming* (New York: Bloomsbury, 2010).

15. Thomas F. Gieryn, "Boundary-Work and the Demarcation of Science from Non-Science: Strains and Interests in Professional Ideologies of Scientists," *American Sociological Review* 48 (1983): 781–795.

CONTRIBUTORS

GARLAND E. ALLEN is Professor of Biology Emeritus at Washington University in St. Louis, former coeditor of the *Journal of the History of Biology,* and past president of the International Society for the History, Philosophy, and Social Studies of Biology. In addition to writing *Life Science in the Twentieth Century* (1975) and *Thomas Hunt Morgan: The Man and His Science* (1978), he has coauthored several introductory biology textbooks.

THEODORE ARABATZIS is Professor of History and Philosophy of Science in the Department of the History and Philosophy of Science at the University of Athens. He is the author of *Representing Electrons: A Biographical Approach to Theoretical Entities* (2006), and coeditor (with Vasso Kindi) of *Kuhn's The Structure of Scientific Revolutions Revisited* (2012).

RICHARD W. BURKHARDT JR. is Professor of History Emeritus at the University of Illinois, Urbana-Champaign. His publications include *The Spirit of System: Lamarck and Evolutionary Biology* (1977) and *Patterns of Behavior: Konrad Lorenz, Niko Tinbergen, and the Founding of Ethology* (2005).

LESLEY B. CORMACK is Professor of History and Dean of the Faculty of Arts at the University of Alberta. She is also president of the Canadian Society for the History and Philosophy of Science, and first vice president of the International Union for the History and Philosophy of Science, Division of History of Science and Technology. She is the author of *Charting an Empire: Geography at Oxford and Cambridge, 1580–1620* (1997) and coauthor of *A History of Science in Society: From Philosophy to Utility* (2004).

DAVID J. DEPEW is Professor of Communication Studies and Rhetoric of Inquiry Emeritus at the University of Iowa. He is coauthor with Bruce H. Weber of *Darwinism Evolving: System Dynamics and the Genealogy of Natural Selection* (1996) and with the late Marjorie Grene of *Philosophy of Biology: An Episodic History* (2004). He is currently working with John P. Jackson on a book tentatively titled *Darwinism, Democracy, and Race in the American Century.*

PATRICIA FARA is the Senior Tutor of Clare College Cambridge and the author of the prize-winning book *Science: A 4000 Year History* (2009), which has been translated into nine languages. Her numerous other publications include *Newton: The Making of Genius* (2002) and *Pandora's Breeches: Women, Science and Power in the Enlightenment* (2004).

KOSTAS GAVROGLU is Professor of History of Science in the Department of the History and Philosophy of Science at the University of Athens. He is the coauthor with Ana Simões of *Neither Physics nor Chemistry: A History of Quantum Chemistry* (2012) and the editor of *The History of Artificial Cold: Scientific, Technological and Cultural Aspects* (2014).

MICHAEL D. GORDIN is Rosengarten Professor of Modern and Contemporary History at Princeton University. He is editor of several volumes and the author of five books, including *A Well-Ordered Thing: Dmitrii Mendeleev and the Shadow of the Periodic Table* (2004), *The Pseudoscience Wars: Immanuel Velikovsky and the Birth of the Modern Fringe* (2012), and *Scientific Babel: How Science was Done Before and After Global English* (2015).

PETER HARRISON is an Australian Laureate Fellow and Research Professor and Director of the Institute for Advanced Studies in the Humanities, University of Queensland, Australia. From 2006 to 2011 he served as the Andreas Idreos Professor of Science and Religion at the University of Oxford. He has published a number of books on the historical relations between science and religion, the most recent of which is *The Territories of Science and Religion* (2015), based on his Gifford Lectures.

JOHN L. HEILBRON is Professor of History and Vice Chancellor Emeritus at the University of California, Berkeley; senior research fellow at Worcester College, Oxford; and visiting professor at Yale University and the California Institute of Technology. He remains active in retirement near Oxford, as evidenced by his recent books, *Galileo* (2010) and, with Finn Aaserud, *Love, Literature and the Quantum Atom: Niels Bohr's Trilogy of 1913 Revisited* (2013).

KOSTAS KAMPOURAKIS is a researcher in science education at the University of Geneva. He is the editor in chief of the international journal *Science & Education* and of the book series Science: Philosophy, History and Education, published by Springer. He is also the author of *Understanding Evolution* (2014) and the editor of *The Philosophy of Biology: A Companion for Educators* (2013).

MICHAEL N. KEAS is Professor of the History and Philosophy of Science at the College at Southwestern in Fort Worth, Texas. He has recently published a number of essays that contribute to his forthcoming book, *Everything Since Kepler,* which explores pivotal episodes and enduring issues in the history and philosophy of science and religion since Johannes Kepler (1571–1630).

ERIKA LORRAINE MILAM is Associate Professor of History at Princeton University. She is author of *Looking for a Few Good Males: Female Choice in Evolutionary Biology* (2011) and coeditor, with Robert A. Nye, of *Scientific Masculinities* (2015).

JULIE NEWELL is Interim Dean of Arts and Sciences at Southern Polytechnic State University, a member of the International Commission on the History of Geological Sciences, and past Chair of the History of Geology Division of the Geological Society of America. Her research focuses on "Geology and the Emergence of Science as a Profession in the United States."

MANSOOR NIAZ is a Professor of Science Education at Universidad de Oriente, Venezuela. He has published over 150 articles in international refereed journals and eight books, including, *Critical Appraisal of*

Physical Science as a Human Enterprise (2009) and *From "Science in the Making" to Understanding the Nature of Science* (2012).

RONALD L. NUMBERS is Hilldale Professor Emeritus of the History of Science and Medicine at University of Wisconsin–Madison and past president of both the History of Science Society and the International Union of the History and Philosophy of Science. He has served as editor of *Isis* (1989–1993) and as coeditor of the eight-volume *Cambridge History of Science*. He has written or edited more than thirty books, including *The Creationists: From Scientific Creationism to Intelligent Design* (expanded ed., 2006) and *Galileo Goes to Jail and Other Myths about Science and Religion* (2009), which has been translated into eight languages.

KATHRYN M. OLESKO is Associate Professor of the History of Science at Georgetown University. She writes on science education, measuring practices, and the development of science and engineering in Prussia in the early modern and modern periods. She is the author of *Physics as a Calling: Discipline and Practice in the Koenigsberg Seminar for Physics* (1991) and the past editor of *Osiris* (12 volumes).

LAWRENCE M. PRINCIPE is Drew Professor of Humanities at Johns Hopkins University, teaches in the Departments of the History of Science and Technology and of Chemistry, and serves as director of the Singleton Center for the Study of Premodern Europe. He specializes in early modern science, particularly alchemy. His numerous publications include *The Scientific Revolution: A Very Short Introduction* (2011) and *The Secrets of Alchemy* (2013).

PETER J. RAMBERG is Professor of the History of Science at Truman State University in Kirksville, Missouri, where he teaches history and philosophy of science and organic chemistry. His research concerns the intellectual and institutional context of chemistry in the nineteenth century. The author of *Chemical Structure, Spatial Arrangement: The Early History of Stereochemistry, 1874–1914* (2003), he is currently writing a biography of the German chemist Johannes Wislicenus (1835–1902).

ROBERT J. RICHARDS is Morris Fishbein Distinguished Service Professor of the History of Science at the University of Chicago, where he is a professor in the Departments of History, Philosophy, and Psychology, and a member of the Committee on Conceptual and Historical Studies of Science. He also directs the Fishbein Center for the History of Science and Medicine. His primary areas of research are German Romanticism and the history of evolutionary theory. He is the author of several books, including *Darwin and the Emergence of Evolutionary Theories of Mind and Behavior* (1987), which won the Pfizer Prize from the History of Science Society; *The Romantic Conception of Life: Science and Philosophy in the Age of Goethe* (2002); and *The Tragic Sense of Life: Ernst Haeckel and the Struggle over Evolutionary Thought* (2008).

DAVID W. RUDGE is Associate Professor with a joint appointment in the Department of Biological Sciences and the Mallinson Institute for Science Education at Western Michigan University. He has written numerous articles—from the perspectives of history, philosophy, and science education—on H. B. D. Kettlewell's classic investigations of natural selection using the peppered moth, *Biston betularia*.

JOHN L. RUDOLPH is Professor of Science Education in the Department of Curriculum and Instruction at the University of Wisconsin–Madison. He also holds affiliate appointments in the Departments of History of Science and Educational Policy Studies. The author of *Scientists in the Classroom: The Cold War Reconstruction of Science Education* (2002), he currently serves as editor in chief of the international journal *Science Education*.

NICOLAAS RUPKE recently retired from the Chair of the History of Science at Göttingen University and now holds the Johnson Professorship of History and Leadership Studies at Washington and Lee University in Virginia. Trained in both geology and the history of science, he has written several books, including *The Great Chain of History: William Buckland and the English School of Geology* (1983), *Richard Owen: Biology without Darwin* (1994), and *Alexander von Humboldt: A Metabiography* (2008). He is currently writing a book about non-Darwinian traditions in evolutionary biology.

MICHAEL RUSE is Lucyle T. Werkmeister Professor of Philosophy and Director of the Program in the History and Philosophy of Science at Florida State University. A prolific author, he has written or edited numerous books, including most recently *The Gaia Hypothesis: Science on a Pagan Planet* (2013) and *The Cambridge Encyclopedia of Darwin and Evolutionary Thought* (2013). With his long-time historical nemesis Robert J. Richards he is coauthoring a book called *Debating Darwin.*

MICHAEL H. SHANK teaches the history of science before Newton at the University of Wisconsin–Madison, where he has been since 1988. He is the author of *"Unless You Believe, You Shall Not Understand:" Logic, University, and Society in Late Medieval Vienna* (1988), the editor of *The Scientific Enterprise in Antiquity and the Middle Ages* (2000), and the coeditor of the second volume of *The Cambridge History of Science: Medieval Science* (2013). He is currently completing a study of the fifteenth-century German astronomer Regiomontanus.

ADAM R. SHAPIRO is Lecturer in Intellectual and Cultural History at Birkbeck–University of London. He is the author of *Trying Biology: The Scopes Trial, Textbooks, and the Antievolution Movement in American Schools* (2013). He is currently working on two forthcoming books: one on "William Paley and the Evolution of the Natural Theology Movement"; the other on a previously unknown pre-Scopes evolution trial, in Nebraska in 1924.

BRUNO J. STRASSER is Professor of the History of Science at the University of Geneva and Adjunct Professor of the History of Medicine at Yale University. He has written numerous articles on the history of medicine and molecular biology, both in history and in science journals. His first book, *La Fabrique d'une Nouvelle Science: La Biologie Moléculaire à l'Age Atomique, 1945–1964* (2006), won the Henry E. Sigerest Prize from the American Association for the History of Medicine. He is currently finishing his second book, on the history of big-data biology.

DANIEL P. THURS graduated with a PhD in the history of science from the University of Wisconsin–Madison. Since that time, he has worked at a number of institutions, including Cornell, Oregon State University, the University of Portland, New York University, and, most recently, the University of Wisconsin–Madison. His first book, *Science Talk:*

Changing Notions of Science in American Culture (2007), explored changing meanings of science during the nineteenth and twentieth centuries. His current research focuses on the rhetorical relationships between science and fear.

INDEX